WATER POLLUTION BIOLOGY

WATER POLLUTION BIOLOGY

A LABORATORY/FIELD HANDBOOK

ROBERT A. COLER, PH.D.
JOHN P. ROCKWOOD, PH.D.

University of Massachusetts at Amherst

TECHNOMIC
PUBLISHING CO., INC.
LANCASTER · BASEL

Water Pollution Biology
a TECHNOMIC® publication

Published in the Western Hemisphere by
Technomic Publishing Company, Inc.
851 New Holland Avenue
Box 3535
Lancaster, Pennsylvania 17604 U.S.A.

Distributed in the Rest of the World by
Technomic Publishing AG

Copyright © 1989 by Technomic Publishing Company, Inc.
All rights reserved

No part of this publication may be reproduced, stored in a
retrieval system, or transmitted, in any form or by any means,
electronic, mechanical, photocopying, recording, or otherwise,
without the prior written permission of the publisher.

Printed in the United States of America
10 9 8 7 6 5 4 3 2 1

Main entry under title:
   Water Pollution Biology: A Laboratory/Field Handbook

A Technomic Publishing Company book
Bibliography: p.

Library of Congress Card No. 89-50810
ISBN No. 87762-655-3

Table of Contents

Preface   vii

1. THE DATA BASE 1
    EXERCISE 1 Eutrophication: A Review of Lake Chemistry   1
    EXERCISE 2 River Diurnal: A Chemical Inventory   4

2. STRUCTURAL ASPECTS OF BIOTIC COMMUNITIES—
BENTHIC MACROINVERTEBRATES AND PERIPHYTON 9
    EXERCISE 3 Benthic Macroinvertebrate Distribution
                     and Sampling Strategies   10
    EXERCISE 4 Diversity   20
    EXERCISE 5 Determination of Diversity with
                     Artificial Substrates   32
    EXERCISE 6 The Application of Diatoms to the
                     Assessment of Water Quality   38
    EXERCISE 7 Algal Chlorophyll Quantitation   43

3. COMMUNITY FUNCTION 49
    EXERCISE 8 Primary Productivity and Photosynthetic Efficiency Using
                     the Light-Dark Bottle Method   49
    EXERCISE 9 A Field Method for Assessing the Water Quality of a
                     Stream Using Primary Productivity   56

4. BIOASSAY AND TOXICITY TESTING 63
    Aquatic Toxicology: Principles and Procedures   63
    EXERCISE 10 Toxicity Testing with Fish   81

EXERCISE 11 Algal Toxicity Testing in a Flow-Through
          Glass Coil Assembly   87
EXERCISE 12 Toxicity Testing with *Daphnia*   92
EXERCISE 13 The Measurement of Dragonfly Respiratory and
          Excretory Rates as Short-Term Indices of Stress   99

Preface

SPORADIC, but significant research in water pollution biology occurred well over a hundred years ago in England, France, and Germany. The chlorination of sewage effluent as a health measure in the early 1900s, however, effectively stymied further growth of the discipline. It wasn't until the 1950s in the U.S.A. that an awakening of environmental awareness caused the shift from a public health to an ecological perspective. Criteria and remedial measures to upgrade water quality necessitated biological input. Early investigators such as C. Tarzwell, P. Doudoroff, and R. Patrick mounted what was essentially a four-pronged effort to create methodologies for the interpretation of the biological response to chemical and physical stimuli: (1) indicator organisms; (2) community structure (diversity, dominance, etc.); (3) community function (respiration and photosynthesis); and (4) bioassay and toxicity tests. Since these avenues of study evolved sequentially, it is our intent to direct the student along the same path.

The methods described here have been generally adopted by consulting firms and state and federal water pollution control agencies. The manual, therefore, is applications oriented and is not designed to serve the limnologist. The chemical and physical parameters of an experimental and control station measured in Chapter 1 provide the basis against which the biological data, retrieved from the same stations, in Chapters 2, 3, and 4 are correlated.

As the student refines his/her laboratory techniques, hopefully he/she will also learn that the "parts per million" perspective is limited. While the chemist is capable of rapidly generating prodigious quantities of data, these parameters attain significance only in a biological context, for pollution is fundamentally a biological phenomenon. The chemical inventory provides a precise "roll call" of what could be impinging on or contributing to the aquatic ecosystem. But the presence of a particular chemical in a specific quantity does not necessarily assure its biological accessibility; it is for the organism to integrate the total stimulus over time. Furthermore, chemical data provide no dimension of time. They only pertain to the sample analyzed.

Probably the most informative data in the assessment of environmental quality are those derived from a competent inventory of the benthic macroinvertebrates. Because of their limited vagility, ubiquitous distribution, and high diversity and abundance, they remain the best indices of water quality. Two areas of expertise must be cultivated before this approach can be implemented. Chapter 2 is devoted to the acquisition of these skills: the identification to the generic level of the major aquatic invertebrate orders and the collection of ecologically and statistically significant samples. The exercises bring students from developing sampling strategies in the laboratory to their application in the field. The retrieved samples (artificial substrates, Surber square-foot samples, etc.) are processed to identify index organisms and derive abundance and diversity values.

Having addressed community composition and structure, the perspective evolves, in Chapter 3, to an investigation of community function. Students will focus on net and primary productivity in lentic and lotic habitats. While previous work centered on the macroinvertebrates, this chapter will be confined to using planktonic and periphytic primary productivity and respiration levels to infer nutrient and toxicant concentrations.

The bioassay protocols delineated in the last chapter (4) provide the most precise answers of all the tools in the pollution biologist's arsenal. However, by virtue of the controlled laboratory environment demanded by this approach, it affords the weakest extrapolation to natural communities. The data are limited to narrow generalizations about the studied life stage of the species tested. The formulation of application factors stipulating limiting concentrations for a whole family of toxicants (i.e., anionic detergents, heavy metals, organophosphate pesticides, etc.) on an entire ecosystem demands great courage. Impressive studies have been undertaken, however, to standardize and refine techniques for extending bioassays to whole life cycles and more than one trophic level. Such avenues of development will be of paramount importance to environmental management.

Thus, to evaluate the impact of stress, the pollution biologist draws upon data from several sources: bioassay, toxicity, productivity, and diversity. Each area feeds into the general fund of knowledge. The problems of time and cost investiture, however, remain critical, for the pollution control agency must justify expenditures in terms of derived benefit. The applied biologist, therefore, unlike the researcher, must compromise. He/she must invest his/her effort wisely and budget his/her time to obtain the fullest possible picture of environmental quality. Because of the impetus provided by federal legislation in the early 1970's mandating the incorporation of an ecological perspective into water surveys, the discipline is enjoying an exponential growth phase. Consequently, the protocols outlined here can only provide a working knowledge into the fundamental approaches of water pollution biology.

The authors would be remiss if we failed to acknowledge the creative contributions of Manuel Correa, Michael Lizotte, Carmen Medeiros, Joel Pratt, Bruce Tease, and David L. Coburn for the cover art.

R. A. Coler
J. P. Rockwood
Amherst
August, 1988

CHAPTER 1

The Data Base

EXERCISE 1/EUTROPHICATION: A REVIEW OF LAKE CHEMISTRY

PURPOSE

THIS introductory exercise is designed to review the basic sequence of lake dynamics that predisposes standing waters to eutrophication as well as to refine your analytical techniques. Accordingly, the chemical milieu characterizing summer stratification and stagnation in a eutrophic lake will be replicated in an illuminated, refrigerated tank (Coler and Romanow, 1975; Visco et al., 1979). While we can produce, in effect, an epilimnion, metalimnion, and hypolimnion, extrapolation to trophogenic and tropholytic zones is not valid, for depth restrictions militate against the attainment of a true compensation level. In this case, photosynthesis is depressed by temperature, rather than by light extinction.

INTRODUCTION

Likens (1972) describes this aging process as the enhanced primary productivity stimulated by organic matter and/or nutrient enrichment resulting in the depreciation of the lake as a recreational resource. Accordingly, we are concerned with those factors that alter the distribution and the consequent biological availability of nitrogen and phosphorous compounds in the trophogenic zone. Probably the most conspicuous chemical factor dictating the form, solubility, circulation, and precipitation of nutrients is the oxidative state of the water column and the underlying sediments of the microzone. For this reason, the class will relate iron, inorganic carbon, nitrogen (NH_3, NO_2^-, and NO_3^-), pH, and phosphorous (ortho and total) levels to metabolic activity. Since primary productivity is almost exclusively a photosynthetic phenomenon, it will be necessary to trace this process at night as well as during the day.

PROCEDURE

These laboratory periods will be used to identify and demonstrate the physical and chemical factors governing nutrient availability.

Reagent Preparation

Each student will be assigned a parameter group for which he/she will assume responsibility. This will extend to the preparation and maintenance of those reagents requisite for its analysis (APHA, 1985). Also, if indicated, he/she should refresh his/her classmates on the analytical procedure and the basic chemistry of the process. As these reagents will be used in both Exercise 1 and 2, enough of each should be prepared to permit a minimum of 50 analyses.

Sampling

To separate the effects of photosynthesis on the distribution and solubilities of the various chemical constituents from those of respiration, collect samples during both the light and dark portions of the light cycle. During each sampling period, carefully siphon a sufficient quantity of water from each layer (epilimnion, metalimnion, and hypolimnion) to permit duplicate analyses for each parameter being measured. Perform your analyses scrupulously using the procedures detailed in APHA (1985). Record your results [mean ($\bar{x}$) and standard deviation (s), see Exercise 3 for calculations] in Table 1-1. Collect the data for the other parameters from the other groups and compare the results with those found in Wetzel (1983) and APHA (1985).

QUESTIONS

(1) Referring to the data (Table 1-1), trace the sequence of events that account for the observed chemical and physical changes with depth and time. In your answer, pay particular attention to DO, CO_2, NO_3^-, iron, and calcium levels. Wetzel (1983) serves as an excellent reference.

(2) Based on these data, what remedial treatment would you inaugurate to halt or reverse the process?
(3) What would be the advantage of initiating a chemical inventory in the spring?
(4) How many stations and what sampling frequency would be required to survey a stream-fed single basin lake with a single outlet?

TABLE 1-1 Concentrations of indicated parameters in the epilimnion, metalimnion, and hypolimnion of a stratified aquarium during day and night sampling (mean and standard deviation of two measurements).

	Day						Night						
	epi-		meta-		hypo-		epi-		meta-		hypo-		
Parameter	$\bar{x}$	s	$\bar{x}$	s	$\bar{x}$	s	$\bar{x}$	s	$\bar{x}$	s	$\bar{x}$	s	Student
pH													
Acidity													
Alkalinity HCO_3^- CO_3^{2-}													
CO_2													
DO													
% Saturation													
BOD													
Ca													
Hardness													
Fe^{2+}													
P_{ortho}													
P_{total}													
NH_3-N													
NO_2-N													
NO_3-N													
Conductivity													
Suspended Solids													
Turbidity													
Temperature													

REFERENCES

APHA. *Standard Methods for the Examination of Water and Wastewater*, 16th ed. American Public Health Association, Washington, D.C., 1268 pp. (1985).

Coler, R. A. and L. Romanow. "An Instructional Model to Demonstrate Thermal, Chemical and Biological Stratification," *AIBS Educ. Rev.*, 4:1-2 (1975).

Likens, G. E. "Eutrophication and Aquatic Ecosystems," in *Nutrients and Eutrophication: The Limiting Nutrient Controversy*. G. E. Likens, ed. American Society of Limnology and Oceanography, Lawrence, Kansas, pp. 3-13 (1972).

Visco, S., R. Coler and O. T. Zajicek. "An Evaluation of Benthic Nutrient Regeneration in a Projected Flood Control Impoundment," *J. Environ. Sci. Health*, A14:399-414 (1979).

Wetzel, R. G. *Limnology*, 2nd ed. Saunders College Publishing, Philadelphia, 767 pp. (1983).

EXERCISE 2/RIVER DIURNAL: A CHEMICAL INVENTORY

PURPOSE

As in the preceding exercise, the intent of this laboratory period is to generate sufficient data to permit comparison of water quality between sampling stations and support speculation regarding their biological carrying capacities. In this instance, we will be generating a data base for a riverine rather than a standing water habitat.

INTRODUCTION

The dominant phenomenon in lake systems is circulation which is responsible for the oxidation and precipitation of nutrients followed by stratification which leads to stagnation, thus causing chemical reduction and their solution. In rivers, except those affected by gross pollution, there is no such sequence. Rather, there is a seasonal pattern of spring erosion followed by deposition. In the river's course to the sea, the process of baseline leveling results in the evolution of mountain riffles and pools to flood plain meanders. The limiting factors, instead of oxygen as in lentic waters, are current and stream bed load. Except for hatched fish, the biota are concentrated in the river bed, whether the food web is grazing or detritus based.

Perhaps of even greater importance is the vulnerability of flowing waters to external inputs. The ratio of shore line to water volume is much greater in streams than in lakes. Given the fact that most pollutants adsorb to suspended particles and slowly leach out upon settling, river chemistry should include analysis of the interstitial and boundary layer waters where the biotic community is concentrated.

Subsequent laboratory exercises will be biologically oriented, focusing on the application of various indices to resolve subtle differences in water quality between the control and test stations studied in this exercise. Remember, it is the aquatic biota that integrate the total environment, thus providing a yardstick for assessing water quality. The biota provide a summation of the water conditions over time, whereas corresponding chemical and physical tests must be performed over a longer period to provide average values. Also, according to Tarzwell (1957), biological surveys and investigations are valuable tools for determining the effectiveness of natural purification processes of streams and their assimilation of and accommodation to changes in the environment by effluent discharges and allochthonous materials.

As analytical values in streams may differ with flow, depth, and distance from the shore, it is best, if the equipment is available, to take an integrated sample, composited by flow, from top to bottom in midstream (Hem, 1970; APHA, 1985). Since the mean velocity of the stream is at a depth approximately 0.6 of the distance from the surface to the bed, it has been suggested that samples be collected from this depth (Klein, 1959), although middepth is often recommended (Kittrel, 1969; APHA, 1985).

PROCEDURE

Since we are occupied here with environmental suitability, our concern is with extremes. Consequently, it is important to monitor the water quality over a 24-hour period to identify batch release regimes often employed by some industries.

For your assigned parameters, use the reagents prepared in Exercise 1 to conduct a preliminary analysis of the stream water to determine concentration ranges. Also, if the study site has a problem with a particular pollutant not listed in Table 2-1, reagents should be prepared and its concentration measured. Prepare to perform duplicate analyses for samples from both stations at four-hour intervals. For our purposes, a single grab sample, of sufficient volume to allow all of the necessary determinations to be made, collected from the middle of the stream at middepth will be adequate for each station for each sampling period. Collect, store, and analyze the samples as described in APHA (1985).

The flow rate should be measured at both sites at the beginning and end of the survey. To do this, determine the cross-sectional area of the stream at both sites. Then, using a current meter, determine the average stream velocity at

TABLE 2-1 **Concentrations of indicated parameters at stations one and two at four-hour intervals (mean of two measurements).**

Parameter	Sampling Period												Student
	Station		Station		Station		Station		Station		Station		
	1	2	1	2	1	2	1	2	1	2	1	2	
pH													
Acidity													
Alkalinity$_{total}$													
CO_2													
DO													
% Saturation													
BOD													
Ca													
Hardness													
P$_{ortho}$													
P$_{total}$													
NH_3–N													
NO_2–N													
NO_3–N													
Conductivity													
Suspended Solids													
Turbidity													
Temperature													
Flow													
Others													

each site. The flow (m^3 or ft^3/second) is equal to the cross-sectional area multiplied by the average stream velocity.

Record the data in Table 2-1. Plot the parameter concentrations versus time for both stations, using a single graph for each group of parameters.

QUESTIONS

(1) Compare the trends evident here with diurnal changes observed in lakes.
(2) How would you correlate changes in water quality in time and distance with land use and human activity? Prepare a land use map from a USGS topographic map to explain mass balance.
(3) To what extent is primary productivity autochthonous? Support your answer with field data.

REFERENCES

APHA. *Standard Methods for the Examination of Water and Wastewater,* 16th ed. American Public Health Association, Washington, D.C., 1268 pp. (1985).

Hem, J. D. *Study and Interpretation of the Chemical Characteristics of Natural Water*, 2nd ed. Geological Survey Water-Supply Paper 1473, U.S. Government Printing Office, Washington, D.C., 363 pp. (1970).

Kittrell, F. W. *A Practical Guide to Water Quality Studies of Streams*. Federal Water Pollution Control Administration, U.S. Government Printing Office, Washington, D.C., 135 pp. (1969).

Klein, L. *River Pollution I. Chemical Analysis*. Butterworth & Co., London, 206 pp. (1959).

Tarzwell, C. M. "Water Quality Criteria for Aquatic Life," in *Biological Problems in Water Pollution*, C. M. Tarzwell, ed. Public Health Service, Cincinnati, Ohio, pp. 246-272 (1957).

CHAPTER 2

Structural Aspects of Biotic Communities – Benthic Macroinvertebrates and Periphyton

PURPOSE

POLLUTION biologists are severely limited by time and cost constraints in their attempt to obtain and process statistically representative samples. Both the taxonomy and contagious distribution of the organisms preclude a rapid assessment of the biotic response. Meaningful chemical data, on the other hand, can be generated on site in a fraction of the time from far fewer samples. Hence, the biological component of a survey is often quantitatively weak—though it is the most critical.

This chapter intends to acquaint you with some of the measures implemented by field biologists, to familiarize you with the major taxa, and to demonstrate their application as indicators of water quality.

INTRODUCTION

The Water Quality Act of 1965 (PL 89-234), an amendment of the Federal Water Pollution Control Act of 1956 (PL 84-660), required each state to establish standards for all interstate waters, including coastal waters, and to develop a plan of implementation and enforcement of these limits. A more recent amendment, the Clean Water Act of 1977 (PL 95-217), established a national goal to prohibit the discharge of toxic substances in toxic amounts. Both pieces of legislation stress the importance of the biota in establishing water quality criteria.

Much emphasis is currently being placed on chemical and physical tests because these parameters, which provide such data as temperature, dissolved oxygen, pH, salinity, hardness, turbidity, etc., are easily defined and produce an estimate of the present water quality. However, such data provide no biological perspective into chronic synergistic effects of subclinical stress.

9

Fish have well-documented life histories and food and habitat preferences, but their motility and low population densities make their distribution difficult to interpret. Microorganisms, on the other hand, respond readily and in astronomical numbers to pollutants through alterations in their community composition. Since the establishment of standard procedures for the routine identification and quantification of fecal streptococcus and coliform bacteria, these organisms have become synonymous with domestic pollution. They would seem to be the perfect choice, but unfortunately, bacteria, as well as protozoa and algae, have limited pertinence to industrial pollution. Also, their short life-cycles and successional sequences do not readily lend themselves to interpretation, and population changes often seem erratic and ephemeral.

Benthic invertebrates, though, are particularly suitable for such studies. Because of their habitat preferences, longer life-cycles, and relatively low motility, they provide a sort of faunal memory of past stresses. However, while the presence of a species indicates that certain minimal environmental requisites have been met, its absence is not as strong an indicator of water quality. Furthermore, relatively large numbers of samples must be collected and analyzed to determine accurate estimates of population numbers and diversity.

Presently, to refine the indicator organism approach, mathematical and statistical techniques are being evolved. Evaluating the biological response in the field has developed from a mere inventory (Exercise 3) to "stream bed" diversity (Exercise 4) to artificial substrate diversity (Exercise 5). Each refinement generates more rapidly and readily reproducible data. Diversity is an important measure of community health and, thus, water quality. These exercises will investigate the abundance and diversity of stream macroinvertebrates—an important link in the food chain.

The last two exercises in this unit will examine differences in the diatom communities of the two sites chosen for study. In Exercise 6, the comparison will be in terms of diversity and generic abundance, while Exercise 7 will examine differences in biomass using chlorophyll extraction techniques.

EXERCISE 3/BENTHIC MACROINVERTEBRATE DISTRIBUTION AND SAMPLING STRATEGIES

PURPOSE

When attempting to assess the impact of a suspected trauma on the benthic macroinvertebrate community, the field worker is confronted with basically two fundamental problems: (1) obtaining representative samples and (2) identifying the biota. This exercise will compare the various sampling strategies using invertebrates collected with a Surber square-foot sampler and identified to family as a means of evaluating the methodologies.

INTRODUCTION

The number of samples required for an estimate of population parameters invariably remains a compromise between logistics and accuracy. Studies demonstrate that at least 194 square-foot samples are required to obtain a 95% confidence value for weight and 73 samples for numbers (Needham and Usinger, 1956). Chutter (1972), in reexamining Needham and Usinger's data, determined that 448 samples would be required to give a sample mean for numbers within 5% of the population mean with a confidence level of 95%. Either of these values, 73 or 448, is clearly an impossible number of samples to collect from the average riffle, let alone process. At best, these quantitative estimates for numbers and biomass are rough.

Needham and Usinger (1956) found that two or three square-foot samples were sufficient to be fairly certain of obtaining the principal taxa. Gaufin et al. (1956) reported that at least eight samples were required to obtain 85–90% of the species, but only three were needed for a 50–65% return. Cairns and Dickson (1971) recommended using at least three samples per site and Chutter and Noble (1966) routinely employed three square-foot samples per habitat type for a faunal survey.

The problem is how to describe and predict benthic macroinvertebrate distribution statistically, when there is neither the resolution to distinguish between microhabitats nor the knowledge to interpret the significance of their differences as they relate to the biota. It is well known that organisms are seldom randomly or evenly distributed, except when density is very low, so that Poisson and positive binomial distribution models are not appropriate. As is most often the case, due to behavioral and habitat requirements, invertebrate populations exhibit a patched or clumped distribution pattern which is best described by the negative binomial model (Elliot, 1977). If the substrate is uniform in composition, we can expect a random distribution of the clumps. Due to this aggregation, a large number of samples is required to describe the community with any degree of confidence.

Because processing biological data is demanding, time and thought should be devoted to the experimental design—except in the case of general surveillance and monitoring where no interpretative studies are required. A simple sampling design (not statistical) using species lists is valid in such cases.

Statistical Techniques

If interpretive data are sought, such as evaluating the impact of various pollutants on the biota, a well-planned experimental design is indicated in which sampling error and variability within samples between similar and different habitats are known. The greater the habitat variability is, both temporally and spatially, the more intensive the sampling regimen should be. We are concerned with the number and size of samples and sampling statistics needed to make valid inferences about the population and to provide a framework of experimental design. Many small samples are preferable (cover a greater variety of habitats, more degrees of freedom) even though sampling errors are magnified. The difficulty is to extract representative data, discarding those due to chance variation and condensing the data to a small, manageable series of numbers.

The process of reducing the data to manageable proportions without doing violence to their substance is the mandate of the statistician. Values which measure variation within a sample are called statistics (Roman letters), and the essential properties of the population they estimate are called parameters (Greek letters). One of the assumptions underlying the use of many statistical procedures is that the data are normally distributed. For sample sizes greater than 30, the normal approximation can be used.

The most commonly used measure of central tendency is the arithmetic mean ($\bar{x}$), which is simply the sum of the individual observations divided by the number of observations.

$$\bar{x} = \frac{\Sigma_i x_i}{n}$$

The sample variance (s^2), a measure of dispersion around the mean, is measured as the sum of the squares of the deviations from the sample mean divided by the number of observations minus one. Because the theoretical formula is cumbersome to use when the number of observations is large, a more practical "working formula" is also provided.

$$s^2 = \frac{\Sigma_i (x_i - \bar{x})^2}{n - 1}$$

or

$$s^2 = \frac{\Sigma_i x_i^2 - [(\Sigma_i x_i)^2 / n]}{n - 1}$$

The standard deviation of a sample (s), a measure of dispersion around the mean in the original scale of measurement, is equal to the square root of the sample variance.

$$s = \sqrt{s^2}$$

A two-statistic summary of the data ($\bar{x} \pm s$) is incomplete unless the sample figures have a normal distribution, and this is rarely the case with this type of biological data (counts). Because the selectivity of the sampling method produces highly skewed samples, more statistics are often required. Transformation of the data (in this case, $\log_{10} x$) often achieves a normal distribution for biological data.

4   10   25   32   63   252   501   631   1000   3990, $\bar{x} = 650.8$, $s = 1220$

0.6   1.0   1.4   1.5   1.8   2.4   2.7   2.8   3.0   3.6, $\bar{x} = 2.08$, $s = 0.97$

In this instance, two statistics would make it possible to reconstruct the figures to approximate the original data. This method is not exact, but it is as good as you could get by returning to the field for a second set of data. These figures then ($\bar{x}$ and s) permit the reduction of hundreds of pieces of data to a mere few. They still remain, though, estimates of the population parameters. Inaccuracy originates from the sampling error and the bias of the sampling technique.

If a sample can be summarized by two statistics, the standard error of the mean ($s_{\bar{x}}$) can be calculated as either the sample standard deviation divided by the square root of the sample size, or the square root of the quotient of the sample variance divided by the sample size. The standard error indicates the amount of error in the sample mean when it is used to estimate the population mean (μ).

$$s_{\bar{x}} = \frac{s}{\sqrt{n}} \quad \text{or} \quad s_{\bar{x}} = \sqrt{s^2/n}$$

If the sample size (n) is 30 or more and the data are not grossly asymmetrical about their mean, the standard error may be used to set upper and lower confidence limits (*CL*) within which the population mean may be assumed to lie with a selected degree of certainty. To determine the confidence interval when data transformation is necessary, see Elliot (1977).

From preliminary sampling, it is possible to calculate the number of samples required to estimate the population mean within some preselected percent error of the sample mean and with a certain probability. This calculation, in addition to those for the previously described statistics, can be found in Examples 1 and 2.

Example 1: Simple Random Sampling

The counts of stonefly nymphs found in eight Surber square-foot samples are indicated below.

Sample	Count
1	12
2	13
3	9
4	7
5	10
6	8
7	10
8	11

The calculations for mean, variance, standard error, and 95% confidence limits are shown below.

$$\bar{x} = \frac{\Sigma_i x_i}{n} = \frac{80}{8} = 10$$

$$s^2 = \frac{\Sigma_i x_i^2 - [(\Sigma_i x_i)^2/n]}{n-1} = \frac{828 - [(80)^2/8]}{7} = 4$$

$$s_{\bar{x}} = \sqrt{s^2/n} = \sqrt{4/8} = 0.707$$

$$CL = \bar{x} \pm (t_{0.05})(s_{\bar{x}}) = 10 \pm (2.365)(0.707) = 10 \pm 1.67$$

The $t_{0.05}$ represents the value of t ($P = 0.05$) where the number of degrees of freedom (df) is equal to the sample size minus one ($n - 1$), in this case, 7. This value can be obtained from the "t distribution" table found in many statistical textbooks.

From the above example, we can say that we are 95% certain that the population mean lies somewhere between 8.33 and 11.67 (or in discrete units, between 8 and 12 individuals per square-foot of substrate).

Suppose now that we want to know the minimum number of samples (N) we need to process to estimate the population mean with a 5% allowable error in the sample mean at the 95% probability level.

$$N = \frac{t^2 s^2}{L^2} = \frac{(2)^2(4)}{(0.5)^2} = 64 \text{ samples}$$

As the number of *df* is unknown, the value of *t* is approximated at 2. This is valid because the value of *t* at the 95% probability level ($P = 0.05$) ranges from 2.042 at 30 *df* to 1.960 at an infinite number of *df*.

L is equal to the product of the sample mean (10) and the preselected, allowable error (0.05); in this case, 0.5 stoneflies. Note that for a 10% error (being within 1 stonefly of the population mean), 16 samples would be required and for a 20% allowable error (2 stoneflies), only 4 samples would be needed.

Example 2: Stratified Sampling

Suppose that the stream bottom (riffle) measures 15 feet by 40 feet and collections are made with a Surber square-foot sampler. Consequently, there are 600 potential sample units ($N = 600$). Suppose further, that the riffle bed is composed of three different substrate types in the indicated proportions: gravel ($N_1 = 120$ units), sand ($N_2 = 300$ units), and silt ($N_3 = 180$ units).

It was decided that only 40 sample units (*n*) from the entire riffle could be processed. The proportional number of sample units (n_i) to be randomly selected from each substrate type is determined as follows.

stratum	$\dfrac{N_i}{N} \times n$	n_i
gravel	$\dfrac{120}{600} \times 40$	8
sand	$\dfrac{300}{600} \times 40$	20
silt	$\dfrac{180}{600} \times 40$	12

To determine the number of mayflies found in each stratum as well as in the riffle as a whole, each stratum was sampled the indicated number of times and the mean and variance determined for each substrate type.

stratum	n_i	$\bar{x}_i$	s_i^2
gravel	8	16	7.14
sand	20	10	5.58
silt	12	9	5.27

The stratified sample mean is calculated by:

$$\bar{x} = \frac{\Sigma_i n_i \bar{x}_i}{n} = \frac{(8)(16) + (20)(10) + (12)(9)}{40} = 10.9$$

Note that a simple mean of the $\bar{x}_i$'s gives a value of 35/3 or 11.7. The standard error of the stratified mean is calculated as:

$$s_{\bar{x}} = \frac{\sqrt{\Sigma_i n_i s_i^2}}{n} = \frac{\sqrt{(8)(7.14) + (20)(5.58) + (12)(5.27)}}{40} = 0.3808$$

If the sampled fraction (n/N) exceeds 0.1, the result from the above equation should be multiplied by the finite correction factor, $\sqrt{1 - (n/N)}$. In either case, the stratified sample variance ($s_{\bar{x}}^2$) is equal to the standard error of the stratified mean squared. The estimated confidence limits for the overall mean of the stratified sample can be calculated using the standard error of the stratified sample mean. In this calculation, 1.960, the value of $t_{0.05}$ for an infinite number of df is used.

$$\bar{x} \pm (t_{0.05})(s_{\bar{x}}) = 10.9 \pm (1.960)(0.3808) = 10.9 \pm 0.75$$

Therefore, the total number of mayflies in the riffle would be expected to be between 6090 and 6990 [600 × (10.9 ± 0.75)]. For more details on stratified sampling and the above calculations, see USGS (1977) and Elliot (1977).

PROCEDURE

There are essentially three sampling methodologies commonly employed by the aquatic biologist: simple random, systematic, and stratified random sampling. The class will be divided into three teams per site (control and test) and each group will be assigned a sampling protocol to estimate the community composition and density in a defined area of a stream bed. The suitability of each team's accumulated data will then be compared at the family level with regard to density as indicated by mean values and confidence intervals.

The stream reach study sites examined in Exercise 2 are 30 feet wide and 60 feet long with each comprised of a riffle and a pool. Using a Surber square-foot sampler as the standard sampling instrument, there are potentially 1800 sampling units per station. Each team will collect six samples from their study site using their assigned sampling strategy.

Regardless of your team assignments, always approach from and work downstream of your sample unit, because virtually any disturbance will provoke a macroinvertebrate to abandon its substrate and drift.

Team 1: Simple Random

Randomly select six sample units from the entire area and collect the specimens accordingly. This method rapidly generates unbiased data; however, it introduces a large element of error because the organisms, as indicated earlier, are usually not randomly distributed.

Team 2: Systematic

The first sample unit should be selected randomly from among the first 300 (1800/6) units and every 300th unit thereafter should be sampled. For example, if 150 was selected as the first unit, the remaining units to be sampled are 450, 750, 1050, 1350, and 1650. This form of sampling often gives more accurate results than simple random sampling because the selected quadrates are more evenly distributed over the entire sampling area.

Team 3: Stratified Random

Divide the site into two sampling areas (riffle and pool) and sample randomly from each. Assign the number of sample units proportionally to each area and collect the specimens accordingly. This method is more useful and accurate because it recognizes uniformities within the community (riffle versus pool) and attempts to sample such that these subdivisions are proportionally represented.

Specimen Analysis

Place the specimens from each sampling unit into separate vials appropriately labeled with team number, sample unit number, and habitat type (riffle or pool) and preserve with 70% ethanol. Process the specimens to family in the laboratory using either Pennak (1978) or Merritt and Cummins (1984) and record the data in Table 3-1.

Compare sampling strategies statistically by calculating the arithmetic mean ($\bar{x}$), variance (s^2), standard error ($s_{\bar{x}}$), and 95% confidence limits (CL) for each family. Record the summary statistics in Table 3-1.

TABLE 3-1 **Family counts and summary statistics.**

Sampling Strategy _____   Site _____

Order	Family	\multicolumn{6}{c	}{Count per Sample Unit and Habitat Type (Pool or Riffle)}	$\bar{x}$	s^2	$s_{\bar{x}}$	CL
		1( )	2( )	3( )	4( )	5( )	6( )				
Ephemeroptera											
Plecoptera											
Trichoptera											
Odonata											
Diptera											
Others											

QUESTIONS

(1) Compare sampling strategies. Which one gives the best return for your effort? Document this.

(2) What can you deduce from the data regarding the dispersal of macroinvertebrates? Would you say that these taxa have a uniform, even, or clumped pattern of distribution? Why?

(3) Which particular group of macroinvertebrates would most likely have a random distribution?

(4) Calculate the mean and variance for the total number of organisms in your six sample units. Using these, calculate the number of samples (N) needed to estimate the population mean with a 5% allowable error in the sample mean at the 95% probability level. How does this value compare with those found in the literature?

REFERENCES

Cairns, J., Jr. and K. L. Dickson. "A Simple Method for the Biological Assessment of the Effects of Waste Discharges on Aquatic Bottom-Dwelling Organisms," *J. Wat. Pollut. Cont. Fed.*, 43:755–772 (1971).

Chutter, F. M. "A Reappraisal of Needham and Usinger's Data on the Variability of a Stream Fauna When Sampled With a Surber Sampler," *Limnol. Oceanogr.*, 17:139–141 (1972).

Chutter, F. M. and R. G. Noble. "The Reliability of a Method of Sampling Stream Invertebrates," *Arch. Hydrobiol*, 62:95–103 (1966).

Elliott, J. M. *Some Methods for the Statistical Analysis of Samples of Benthic Invertebrates*, 2nd ed. Freshwater Biological Association Scientific Publication No. 25, 160 pp. (1977).

Gaufin, A. R., E. K. Harris and H. J. Walter. "A Statistical Evaluation of Stream Bottom Sampling Data Obtained from Three Standard Samplers," *Ecology*, 37:643–648 (1956).

Merritt, R. W. and K. W. Cummins. *An Introduction to the Aquatic Insects of North America*, 2nd ed. Kendall/Hunt Publishing Co., Dubuque, Iowa, 722 pp. (1984).

Needham, P. R. and R. L. Usinger. "Variability in the Macrofauna of a Single Riffle in Prosser Creek, California, As Indicated by the Surber Sampler," *Hilgardia*, 24:383–409 (1956).

Pennak, R. W. *Fresh-Water Invertebrates of the United States*, 2nd ed. John Wiley & Sons, New York, 803 pp. (1978).

USGS. "Methods for Collection and Analysis of Aquatic Biological and Microbiological Samples," *Techniques of Water–Resources Investigations of the United States Geological Survey*, Book 5, Chapter A4, U.S. Dept. of the Interior, 332 pp. (1977).

EXERCISE 4/DIVERSITY

PURPOSE

Having experienced the frustrations of trying to determine community composition (with any degree of confidence) from Surber square-foot samples, we are now prepared to explore the alternatives. The purpose of this exercise is to familiarize you with some of the techniques and tactics which have evolved to conduct statistically valid assessments of natural communities, from the formulation of effective sampling strategies to the interpretation of data.

INTRODUCTION

While the presence of an organism in an environment indicates that its minimal environmental requirements are being met, its absence could be attributed not only to unsuitable environmental conditions, but to lack of access to the area or competition for its particular niche. These common-sense observations form the rationale behind the use of organisms as indicators of pollution. For example, the absence of stoneflies, found in clean, flowing, well-aerated water would tend to indicate organic pollution. The presence of sewage worms in significant numbers would reinforce this hypothesis. The more one knows about the environmental requirements of specific organisms and the more thorough one's collection techniques are, the better is one's insight into past and present physical and chemical conditions.

The type, number, and distribution of organisms in an aquatic habitat are the basic components of community structure, and these reflect the environmental conditions of the life support system. Although biological studies are most faithfully expressed as long lists of types and numbers of species found, this format has little meaning to the layperson, and more importantly, demands a prodigious investment of time and personnel. Consequently, methods have been evolved to condense and express these results numerically (Wilhm, 1972). The diversity index is a function of both species richness and species evenness. Richness or variety pertains to the number of different species comprising a community, while evenness or equitability describes the extent to which each species is represented within the community.

The more diverse a community is, the greater the degree of interaction within and between the biota and the abiotic environment. In a healthy system, species populations are kept in check by the stabilizing forces of predation, parasitism, and competition. Although species composition may change, the diversity will remain fairly constant.

Diversity

Energy and nutrients are constantly being introduced, taken up, released, and taken up again as they cycle through the biotic community. Because of this, the diversity of organisms is intricately related to the chemical composition of a system. In a healthy system, the dissolved oxygen is abundant and nutrient levels neither excessive nor limiting.

The purpose of measuring a community's diversity is usually to judge its relationship either to other community properties, such as productivity and stability, or to the environmental conditions to which the community is exposed. If the only stresses on the system are those chemical, physical, and biological changes which occur in response to the natural seasonal procession, equilibrium will be maintained.

Should an unnatural stress be exerted, for instance, in the form of organic enrichment from a sewage outfall, or thermal pollution from a power plant, evenness will decrease. The organisms which are most sensitive to the stress will decrease in number or be eliminated. The equilibrium maintained by the stabilizing forces of predation and competition will be upset, and those organisms most tolerant to the pollution will increase in number. As a result, the stability and diversity of the system will generally decrease.

Diversity may be considered as the uncertainty faced by an organism in attempting to find another of its own kind, that is, the uncertainty of an intraspecific encounter. This uncertainty increases as the number of species in the community increases and as their proportions become more even. Conversely, if some species become rare or even disappear and a few become dominant, diversity and the average degree of uncertainty will decline. One should note that interspecific encounters (e.g., predator/prey interactions) become more likely as diversity increases. The implications of this are fundamental to the structure and function of biotic communities.

To determine the diversity of a community or collection small enough to completely census, the Brillouin Index is commonly used. The Brillouin Index may be stated as:

$$H = \frac{c}{N}\left(\log_{10} N! - \sum_{i=1}^{s} \log_{10} N_i! \right)$$

where,

H = the diversity per individual in the collection
N = the total number of individuals in the collection
N_i = the total number of individuals in the ith species

S = the number of species in the collection
$i = 1, 2, 3, \ldots, S$
$c = 3.3219$, the constant for converting logarithms of base 10 to base 2

A table of $\log_{10} n!$ can be found in Lloyd et al. (1968) and Weber (1973).

If a community or collection is too large to be completely enumerated, its diversity can be estimated from a randomly selected portion of the collection (Pratt and Coler, 1976). Thus, if the diversity of flora in a meadow of one-thousand square feet were to be determined, ten individual randomly selected square foot patches could be examined, and the information collected used to characterize the diversity of the meadow. In this case, each individual square foot of area could be considered a sample unit (all ten sample units make up the sample) and the diversity index for each sample unit (SU) may be calculated as:

$$h = \frac{c}{n}\left(\log_{10} n! - \sum_{i=1}^{s} \log_{10} n_i!\right)$$

where,

h = the diversity per individual in the sample unit
n = the total number of individuals in the sample unit
n_i = number of individuals in the ith species in a sample unit
s = the number of species in the sample unit
$i = 1, 2, 3, \ldots, s$
$c = 3.3219$

Data from two or more sample units may be combined to form a pooled sample. It has been found that a plot of the pooled diversity index (H_k) versus cumulative sample units becomes asymptotic, and that once this asymptotic value for H_k has been found, little is gained by additional sampling. Thus, calculation and plotting of H_k as each new sample unit is added to the pooled sample will indicate when sampling may be stopped. The pooled diversity index is calculated by:

$$H_k = \frac{c}{N_k}\left(\log_{10} N_k! - \sum_{i=1}^{S_k} \log_{10} N_{ik}!\right)$$

where,

H_k = the diversity per individual in k pooled sample units drawn randomly from a collection or community
N_k = the total number of individuals in the pooled sample
N_{ik} = the cumulative number of individuals in the ith species in the pooled sample
S_k = the total number of distinct species in the pooled sample
$i = 1, 2, 3, \ldots, S_k$
$c = 3.3219$

Evenness (J) is a measure of the relative frequency with which organisms are distributed among the various taxa comprising a community, collection, or sample. J may be estimated by:

$$J \cong \hat{J} = \frac{\hat{H}}{c(\log_{10} S_k)}$$

where,

J = evenness
$\hat{J}$ = an approximation of J
$\hat{H}$ = an approximation of H
S_k = the total number of distinct species in the pooled sample
$c = 3.3219$

For individual sample units, the estimate of evenness (j) should be determined using the same general formula, substituting j, h, and s for $\hat{J}$, $\hat{H}$, and S_k, respectively.

Maximum evenness exists when N is distributed equally among the S species present. Minimum evenness occurs when every species but one is represented by a single individual and the remaining organisms belong to just one species.

PROCEDURE

In this laboratory exercise, a community of organisms will be simulated by means of colored beads (approximately 1000) scattered non-randomly over a sample board (54 four-inch squares) representing the habitat. This simulation allows us to focus on several problems inherent to community studies—sampling (how and how much), data processing and analysis, interpretation of the results, and evaluation of assumption and method—without undertaking actual field work.

Sampling

Using either simple random sampling, stratified random sampling, or systematic sampling, generate sampling data which give an accurate representation of the collection of beads "inhabiting" your sample board. When selecting a sampling strategy, take into account the patchy distribution of the beads—some kinds are clumped together and others are more or less randomly or evenly distributed. Also remember that for sampling to be efficient, it should provide an accurate and reliable picture of the parent community with a minimum of superfluous effort. This is greatly appreciated by a fieldworker on a cold, wet day and by the project administrator on a limited budget.

In order to select the quadrates expeditiously, depending upon which sampling strategy you pick, use either numbered slips of paper (1 to 54) or the numbered borders of the sample board in conjunction with slips of paper with coordinate numbers. The contents of a quadrate constitute a sample unit (SU).

Identification of "Species"

While one person on your team collects the beads from a particular quadrate, another should "identify" them using the provided key. In our laboratory, the key lists twenty-five different "species" or types of beads, each with numerical names, but not all may occur on your board.

Organization of Data

As you sort through the first SU, record on the Raw Data Sheet (Table 4-1) the types and numbers of beads you find. In the column headings are the SU numbers, while the rows list "species" types. Record the number of individuals (n_i) of each "species" type found in each sample unit under the appropriate SU column heading. For instance, if the first SU contains five beads of "species" type 2, then 5 (n_2) should be entered on the second line of the first column. After processing all of the individuals found in the first quadrate, figure the total number of beads identified (n) and the total number of bead types present (s). Then, calculate the diversity (h) and the evenness (j) using the appropriate equations. As indicated earlier, a table of $\log_{10} n!$ can be

TABLE 4-1 Raw Data Sheet.

Species	Sample Unit (SU)														
	1	2	3	4	5	6	7	8	9	10	11	12	13	14
1														
2														
3														
4														
5														
6														
7														
8														
9														
10														
11														
12														
13														
14														
15														
16														
17														
18														
19														
20														
21														
22														
23														
24														
25														
n														
s														
h														
j														

found in Lloyd et al. (1968) and Weber (1973). Record the values for n, s, h, and j in the first SU column on the Raw Data Sheet.

Census the second SU the same way, then pool the data from the first two SUs and record them on the Pooled Data Sheet (Table 4-2). This is best accomplished by adding across the columns for the individual "species" on the Raw Data Sheet. Determine the total number of individuals (N_k) and the number of different bead types (S_k) in the combined sample. Then, using the equation for H_k, calculate the diversity of the first two pooled SUs. Enter these values in the appropriate spaces on the Pooled Data Sheet.

Continue processing successive SUs as outlined above one at a time, and continue to pool the data until either four consecutive "asymptotic" H_k values are attained that differ by less than 0.2 or until 14 SUs have been processed.

Estimate the total number (N) of beads originally on your sampling board. Enter your estimate(s) ($\hat{N}$) in Table 4-3 and indicate the method(s) used to obtain it.

Sampling Efficiency

Enter the appropriate data in Table 4-4 and evaluate your sampling efficiency, i.e., the amount of information gained per SU, by plotting the SU number versus new "species" found. Alternatively, you may construct a graph of either pooled samples (k) or pooled individuals (N_k) versus cumulative "species" (S_k). What does the resulting curve tell you about your sampling efficiency, particularly in the context of field sampling?

Rank-Abundance

Prepare a rank-abundance list of all bead types found in m SUs, where m is the total number of SUs combined to achieve four asymptotic H_ks. To do this, simply order the N_{ik} values from the largest to the smallest and record them in Table 4-5. Construct a graph with "species" rank on the x axis and the number per "species" (N_{ik}) on the y axis. Briefly describe the pattern of numerical abundance observed, that is, what is the basic shape of the curve? Relate this graph to the concept of evenness. Does high, moderate, or low evenness seem to exist in your collection?

Diversity and Evenness

Estimate the diversity and evenness of your collection as follows. Average the individual diversity (h) and evenness (j) values of the m SUs and record

TABLE 4-2 Pooled Data Sheet.

Species	Pooled Sample Units (k)											
	2	3	4	5	6	7	8	9	10	11	12	13	14
1													
2													
3													
4													
5													
6													
7													
8													
9													
10													
11													
12													
13													
14													
15													
16													
17													
18													
19													
20													
21													
22													
23													
24													
25													
N_k													
S_k													
H_k													

TABLE 4-3 Estimation of collection size (*N*).

Sample Unit	Sample Size (*n*)	Sample Unit	Sample Size (*n*)
1		8	
2		9	
3		10	
4		11	
5		12	
6		13	
7		14	

Estimated Collection Size ($\hat{N}$):

Method(s) Used to Estimate *N*:

TABLE 4-4 Sampling efficiency.

Sample Unit	Number of New "Species" Found	Pooled Sample Units (*k*)	S_k	N_k
1				
2		2		
3		3		
4		4		
5		5		
6		6		
7		7		
8		8		
9		9		
10		10		
11		11		
12		12		
13		13		
14		14		

TABLE 4-5 Ranked-abundance.

"Species" Rank	"Species" Name	N_{ik}
1		
2		
3		
4		
5		
6		
7		
8		
9		
10		
11		
12		
13		
14		
15		
16		
17		
18		
19		
20		
21		
22		
23		
24		
25		

TABLE 4-6 **Estimation of diversity.**

Sample Unit	Sample Unit Diversity (h)	Sample Unit Evenness (j)	Pooled Sample Units (k)	Pooled Sample Unit Diversity (H_k)
1				
2			2	
3			3	
4			4	
5			5	
6			6	
7			7	
8			8	
9			9	
10			10	
11			11	
12			12	
13			13	
14			14	

Estimated Diversity ($\bar{h}$):	
Estimated Evenness ($\bar{j}$):	
Estimated Diversity ($\hat{H}$, last of 4 asymptotic H_k values):	
Estimated Evenness ($\hat{J}$):	
Compare $\bar{h}$ and $\hat{H}$ as estimates of H. Which do you trust and why?	

Actual Diversity (H):	Actual Evenness (J):
Percent Error in $\bar{h}$:	and $\hat{H}$:
Percent Error in $\bar{j}$:	and $\hat{J}$:

Note: Percent Error = $\dfrac{\text{actual value} - \text{estimated value}}{\text{actual value}} \times 100$

(Table 4-6) their means as estimates of H and J, respectively. Why might you be dubious of the accuracy of these estimates?

Take the last of the four asymptotic diversities (H_k) as an estimate of H ($\hat{H}$). Calculate $\hat{J}$ using the equation provided earlier and record these values in the appropriate spaces in Table 4-6. Plot k versus H_k to obtain a diversity curve for the pooled data. This should demonstrate that once a certain sample size is reached, diversity, as measured by the Brillouin Index, is relatively independent of sample size. Why should this quality be desirable?

Evaluation of Sampling Strategies

Ask the instructor for your collection's actual size (N), "species" number (S), diversity (H), and evenness (J). Using Tables 4-6 and 4-7, compare these actual values with your own corresponding estimates (i.e., with $\hat{N}$, S_k, $\hat{H}$, $\bar{h}$, $\hat{J}$, and $\bar{j}$) by calculating a percent error for each comparison. Also, what percentage of your collection (N) did you have to census to achieve a pooled diversity value (H_k) that was within 20%, 10%, and 5% (if possible) of your collection's actual diversity (H)?

By following class/laboratory meeting, be prepared to discuss the relationships among "species" richness, evenness, diversity, and sampling effort. Also consider how this exercise might influence the manner in which you would conduct actual field work.

QUESTIONS

(1) Why not use biomass and secondary productivity as well as diversity as indices of environmental quality?

TABLE 4-7 Summary statistics.

Collection and Sampling Method Used	Sample Size		"Species" Richness		Diversity		Evenness		% of N to Give Estimate of H Within		
	N	$\hat{N}$	S	S_k	$\bar{h}$	$\hat{H}$	J	$\hat{J}$	20%	10%	5%

(2) Which habitat would most likely have a higher diversity, an environment that is acutely toxic and stable or one that is almost as toxic but only on an intermittent (weekly) basis? Explain.

(3) Can you think of reasons other than as an index of environmental quality, why the scientific community is concerned with maintaining diversity on a global scale?

REFERENCES

Lloyd, M., J. H. Zar and J. R. Karr. "On the Calculation of Information-Theoretical Measures of Diversity," *Am. Mid. Natur.*, 79:257–272 (1968).

Pratt, J. M. and R. A. Coler. "A Procedure for the Routine Biological Evaluation of Urban Runoff in Small Rivers," *Water Res.*, 10:1019–1025 (1976).

Weber, C. I. *Biological Field and Laboratory Methods for Measuring the Quality of Surface Waters and Effluents*. Environmental Monitoring Series, U.S. Environmental Protection Agency, Cincinnati, Ohio (1973).

Wilhm, J. "Graphic and Mathematical Analyses of Biotic Communities in Polluted Streams," *Ann. Rev. Entomol.*, 17:223–252 (1972).

EXERCISE 5/DETERMINATION OF DIVERSITY WITH ARTIFICIAL SUBSTRATES

PURPOSE

Having been provided, in the diversity index, with an alternative to determining species density, you will apply this approach to communities colonizing artificial substrates. By adopting this tactic, the investigator eliminates sampling error due to variability in the nature and area of the substrate. Differences in resident communities of substrates maintained at two different stations would be due solely to water quality and chance. Because the recruitment area remains undefined, however, this protocol only supports comparisons among substrates and not with river bed communities.

INTRODUCTION

This exercise will be devoted to the application of two methodologies utilizing artificial substrates. The first technique involves the placement of Hester-Dendy substrates (Hester and Dendy, 1962) at each site to assess the relative water quality by comparing diversity values and by determining collection dissimilarity. The second procedure, a modification of the first, serves to determine if the organisms are absent simply due to the inability to reach the substrate or because they are unable to tolerate the water quality (Medeiros et al., 1983). To distinguish between these two possibilities, substrates will be placed at both a control and at the test site for colonization. Substrates from the control site will then be netted and transported to the test site, thus functioning as live traps. Before the transfer, all diversity values for the samplers at the control site are assumed to be the same since the substrates were colonized in comparable environments. This is to say, the variability observed between the test and control site substrates must be greater than that observed within the control substrate communities.

PROCEDURE

Field Methodology

It is important to note that at every stage of the method, care must be taken to assure that each of the samplers is processed in the same manner. Uniform handling will minimize sources of error which could bias the results. All variables with the exception of chemical and physical differences in water quality must be kept constant. The two study riffles identified in earlier exercises as being physically similar (dimensions, flow regime, insolation, temperature, and suspended load) will be compared regarding substrate diversity.

Site Preparation

At each riffle head, drive metal or wooden stakes into each side of the stream bed and connect with a thick (0.5 to 0.7 cm diameter) nylon cord stretched approximately 15 cm above the stream at its lowest point. Prior to installing the upstream and downstream cords, knot 9 and 6 loops spaced equidistantly along their respective lengths.

At the upstream site secure samplers numbered 1 to 9. To reduce siltation and forestall exposure to air, respectively, position the substrates 3 to 5 cm above the stream bed in at least 30 cm of water. Substrates 3, 5, and 9 will provide an indication of variability within the upstream control station, while 2, 4, and 8 and 1, 6, and 7 are designated for netting and deployment at the upstream and downstream stations, respectively. Substrates 10, 11, and 12 should be suspended, in a similar manner, at the downstream test station. Five weeks should be allowed for colonization of the substrates.

Recruitment

Net the six Hester-Dendy substrates designated for recruitment from the upstream site. Macroinvertebrates readily disengage from substrates when disturbed. Therefore, approach the samplers carefully from downstream and quickly slip a netting bag over the substrate and secure it. Immediately place the netted sampler into a bucket of stream water for transport. Carefully suspend the appropriate three samplers at each site as previously described.

This procedure exposes all the samplers to the same amount of stress. At the test site, some macroinvertebrates will die due to changes in water quality, but these will be mixed with those that have died simply from the stress of transport. The netted control substrates will permit you to distinguish between the two causes of death.

Following two additional weeks of exposure, retrieve all substrates and derive their respective diversity values and confidence limits. The non-netted substrates should be carefully placed into individual ziplock bags and kept moist, while the netted samplers may be transported to the laboratory in buckets of stream water from their respective sites.

Laboratory Methodology

Separation

NON-NETTED SAMPLERS

Disassemble the Hester-Dendy substrate in a large wash basin and remove the macroinvertebrates from the surfaces with a soft brush. Concentrate the sample in a No. 30 sieve and remove large specimens for preservation. Take subsamples of the remaining material in the sieve and examine for small specimens using a shallow, white enamel pan. It is important that each sampler be cleansed separately so that collections will not be mixed. Store all specimens for identification in 70% ethanol, making sure the vial is labeled with the substrate number, site, and your name.

NETTED SAMPLERS

The netted samplers should be processed in the same manner as the non-netted samplers except that dead and living specimens should be placed into separate vials. In this way, an accurate measurement of relative carrying capacities can be made as well as a comparison between netted and non-netted techniques. It is extremely important to label the vials accurately.

Identification of Specimens

After the separation procedure is completed, identify the respective families of macroinvertebrates using either Pennak (1978) or Merritt and Cummins (1984).

CALCULATIONS

Diversity

The diversity of the samples will be calculated using Brillouin's Index. The artificial substrates do not represent random samples of the community but rather, selective collections. Therefore, each collection is treated as a finite population for which a complete census is taken. For the netted and non-netted substrates, calculate and record (Table 5-1) the diversity (H) for each substrate using the appropriate formula from Exercise 4. A comparison of the

TABLE 5-1 Diversity values (H) for non-netted and netted samplers.

Substrate	Site	Sampler No.	Diversity	Student
Non-Netted	Control	3		
		5		
		9		
		$\bar{x}$		
	Test	10		
		11		
		12		
		$\bar{x}$		
Netted	Control	2		
		4		
		8		
		$\bar{x}$		
	Test	1		
		6		
		7		
		$\bar{x}$		

netted and non-netted values will determine whether the extra effort of recruiting and netting provided a better indication of water quality.

Percent Dissimilarity

Using the data from the non-netted collections, determine the family, collection, and percent dissimilarity values of the downstream test station (Pratt et al., 1981). Compile a ranked-abundance list of the most common families for both the control and the test sites until 60% of each collection is accounted for (Table 5-2). If a family appears in only one of the collections, its absence in the other should be indicated with a zero. Adjust the raw data by multiplying each individual test site family abundance (N_{Ti}) by the ratio of the

TABLE 5-2 Ranked-abundance list of the control and test site families for the determination of family dissimilarity (*FD*), collection dissimilarity, and percent dissimilarity.

Family	Control N_C	Test N_T	N'_T	AD	FD
1					
2					
3					
4					
5					
6					
7					
8					
9					
10					
11					
12					
13					
14					
15					
Collection Dissimilarity:					
Percent Dissimilarity:					

total number of individuals listed in the control site collection to those listed in the test site collection ($\Sigma N_{Ci}/\Sigma N_{Ti}$) and record this value as the adjusted test site abundance (N'_{Ti}). Determine the family dissimilarity value for each individual listed family by:

$$FD_i = \frac{AD_i}{(N_{Ci} + N'_{Ti})} + \frac{AD_i}{\Sigma N_{Ci}}$$

where,

FD_i = the family dissimilarity of the ith family
N_{Ci} = the observed control site abundance of the ith family
N'_{Ti} = the adjusted test site abundance for the ith family
AD_i = $|N_{Ci} - N'_{Ti}|$, the absolute value of the difference between N_{Ci} and N'_{Ti}
$\Sigma N_{Ci} = N_{C1} + N_{C2} + \ldots + N_{Cf}$
$i = 1, 2, 3, \ldots, f$ listed families

Average the complete series of FDs to obtain the collection dissimilarity (CD). Determine the percent dissimilarity (PD) by:

$$PD = \frac{CD}{1 + (2/f)} \times 100$$

where,

f = the number of listed families

QUESTIONS

(1) Would you expect to find a more diverse community in a habitat whose substrate consists of rock, or in one whose substrate consists of rock and mud? Explain.
(2) Diversity in stressed environments varies considerably throughout the year, but in unstressed environments it is remarkably constant. Explain.
(3) What would be the advantages and disadvantages of basing macroinvertebrate diversity on feeding style, i.e., collectors, grazers, predators, and shredders?

REFERENCES

Hester, F. E. and J. S. Dendy. "A Multiple-Plate Sampler for Aquatic Macroinvertebrates," *Trans. Am. Fish. Soc.*, 91:420–421 (1962).

Medeiros, C., R. LeBlanc and R. A. Coler. "An *In Situ* Assessment of the Acute Toxicity of Urban Runoff to Benthic Macroinvertebrates," *Environ. Toxicol. Chem.*, 2:119–126 (1983).

Merritt, R. W. and K. W. Cummins. *An Introduction to the Aquatic Insects of North America*, 2nd ed. Kendall/Hunt Publishing Co., Dubuque, Iowa, 722 pp. (1984).

Pennak, R. W. *Fresh-Water Invertebrates of the United States*, 2nd ed. John Wiley & Sons, New York, 803 pp. (1978).

Pratt, J. M., R. A. Coler and P. J. Godfrey. "Ecological Effects of Urban Stormwater Runoff on Benthic Macroinvertebrates Inhabiting the Green River, Massachusetts," *Hydrobiologia*, 83:29–41 (1981).

EXERCISE 6/THE APPLICATION OF DIATOMS TO THE ASSESSMENT OF WATER QUALITY

PURPOSE

Diatoms are frequently the dominant microscopic photosynthetic organisms in natural waters. Because most of their nutrition is derived from dissolved nutrients and reduced carbon, they may be used as indicators of prevailing environmental conditions. Two methods will be used to study the relationship between diatoms and water quality. First, the Sequential Comparison Index (SCI) will be employed to determine the diversity of diatom species at various sites in the segment of water under study. Second, in Exercise 7, the biomass will be determined for the same sites using a chlorophyll extraction technique. By comparing the data obtained at each site with these two techniques, as well as with those from previous exercises, an assessment of the water quality can be implemented.

INTRODUCTION

Diatoms are a clearly defined group of algae that can be distinguished by the presence of a siliceous outer cell wall or frustule. They are microscopic and usually unicellular, but occasionally occur in small aggregations. The frustules exhibit an astounding array of shapes and sizes which make microscopic examination very interesting. Diatoms dominate both the plankton and periphyton. They are often the most abundant photosynthetic autotrophs existing in lakes and streams and, as such, the dominant primary producers at the base of the grazing food web. Hence, data describing their distribution and abundance are particularly significant to the pollution biologist.

Although diatoms reflect the chemical characteristics of the water, there are many physical factors such as current speed, temperature, light, and season that affect abundance and species differences. They occur throughout the year, but seem to thrive during the spring and fall in temperate latitudes. Care must be taken to consider all these factors when comparing diatom communities from different areas.

Through careful analysis of water quality and community composition, it was found that the presence of certain species could be used to define various zones of organic pollution. These zones are classified as polysaprobic, mesosaprobic, and oligosaprobic (Kolkwitz and Marsson, 1908). The polysaprobic zone is characterized by a large amount of decomposable organic matter. Chemicals are present in the reduced state and dissolved oxygen is low. The mesosaprobic zone represents the area of recovery from heavy organic pollution in which organic matter is being oxidized, and the amount of dissolved oxygen is increasing. In the oligosaprobic zone, most or all of the organic matter has been mineralized and the dissolved oxygen is high. Each of these zones contains indicator organisms that can be specific for that zone. However, while there are some species that occur only in heavily polluted waters (saprobiontic), there are others that generally occur in polluted waters but may exist elsewhere (saprophilous), some that occur in clean water but can tolerate some pollution (saproxenous), and some that are unable to tolerate any pollution (saprophobous) (Fjerdingstad, 1962).

Although this system is useful, its application is restricted to domestic sewage. Further, it was realized that the species composition in a body of water will vary greatly over time with no apparent change in water quality. Another system was needed to take these natural changes in species composition into account. Patrick et al. (1954) demonstrated that diatoms fit the distributional patterns enunciated by Preston (1948). In a healthy body of water the algal community is represented by a high number of species, each of which contains a small number of individuals (Patrick, 1962). The effect of pollution is to reduce the number of species and increase the number of individuals in the remaining tolerant species. In other words, in the presence of pollution the total diversity of the algal community is lowered though productivity may be increased.

Because this system of measuring pollution depends not on the particular species present, but on the diversity of species, it effectively takes into account natural changes in species composition and is a good indicator of water quality. Many methods have been developed to statistically measure diversity; however, almost all of them depend upon identification to the species level, which is time-consuming and requires a great deal of taxonomic skill. By far, the easiest and fastest method to use is the Sequential Comparison Index developed by Cairns et al. (1968) and expanded by Cairns and Dickson (1971). The organisms are distinguished solely by shape, size, and color, so little or no taxonomic training is necessary. Cairns and Dickson (1971) emphasized that this procedure is not meant to replace the more accurate and complex diversity indices evolved by others, but rather should be used as a simple preliminary method of screening water quality.

PROCEDURE

Collection

The algae will be collected through the use of a diatometer (Patrick et al., 1954). A series of studies carried out at the Academy of Natural Sciences showed that communities which developed on the glass slides of the diatometer were very similar to those found on natural substrates and that only living organisms adhered to the glass. The main asset of slides is that their uniformity makes them the ideal substrate for relative diversity studies, in which the only variable between sites should be the physical and chemical water quality. Sites should be chosen where flow rates and insolation are as similar as possible, since these factors play an important role in controlling diatom adhesion and growth.

Exposure of Slides

Attach the diatometer to a tree, rock, or, if necessary, to a metal or wooden stake driven into the stream bed at the sites designated in Exercise 5. Allow the diatometer to float freely in the water for 12–14 days to provide sufficient time for colonization. Following the colonization period, remove one-half of the slides (at least one per student per site) for use in estimating diversity. The remaining slides will be used to measure biomass in Exercise 7. Transport the slides back to the laboratory in a slide box to avoid abrasion.

Preparation of Permanent Slides

There are two methods of preparing slides for use in determining the diversity of the diatom community via the SCI. The goal of both methods is to produce slides with an optimum concentration of diatoms to expedite the counting procedure.

Method 1

Using a toothbrush, razor blade, or rubber policeman, scrape the periphyton off of the slide and disperse them in distilled water in a centrifuge tube. Centrifuge for roughly 20 minutes at 3000 rpm and make several dilutions of the concentrate. Place a drop or two of the diluted concentrate on several cover slips and place on a hotplate, heating slowly until all of the water has evaporated. If required, the cells can be cleared by igniting the cover slips in a muffle furnace for one hour at 500°C. The siliceous frustules will not be damaged by this process. After the slips have cooled, place several microscope slides, each with one drop of a mounting solution (e.g. Permount), on the hotplate and heat for a few minutes. Invert the cover slips (concentrate side down) on the heated media, one to a slide. Apply slight pressure to each cover slip to remove any bubbles. Remove the slides from the hotplate and allow them to cool.

Method 2

Scrape the periphyton from one side of the slide, leaving the other side intact. Bake the slide in a muffle furnace for one hour at 500°C to drive off the organic material. Allow it to cool and place a drop of mounting solution on the center of the slide. As before, heat slowly for a few minutes on a hotplate, position a cover slip, and apply slight pressure to remove any bubbles. Remove the slide from the hotplate and allow it to cool.

Using the SCI

The slide should be examined under the oil immersion lens (970 to 1000 x). Start at one side of the field of view and begin to count the individual organisms (counting is easier using an occular with a grid). Assign the first organism an x. If the next organisms is the same, record another x; if it is different, record an o. The last organism is then compared to the next, using the same procedure. In this way, only two organisms are compared at once. This procedure is continued sequentially, recording a stream of x's and o's. For example, if the organisms viewed are in this order:

$$AA\ BB\ A\ C\ B\ DDD\ E\ FF$$

then you would record:

$$xx\ oo\ x\ o\ x\ ooo\ x\ oo$$

In this case, there are 13 organisms and 8 runs. To allow a reasonable estimate of diversity to be made, between 200 and 250 organisms should be compared. While counting, the number of different taxa (organism types) observed in the sample should be recorded. The diversity index value (DI_1) is determined by dividing the number of runs by the total number of organisms counted:

$$DI_1 = \frac{\text{runs}}{\text{total count}}$$

The final index value (DI_T) is determined by multiplying DI_1 by the number of different taxa (number of different organism types observed):

$$DI_T = DI_1 \times (\text{number of taxa})$$

In the example:

$$DI_1 = \frac{8 \text{ runs}}{13 \text{ organisms}} = 0.62$$

$$DI_T = 0.62 \times 6 \text{ taxa (A, B, C, D, E, and F)}$$

$$DI_T = 3.72$$

In previous studies, DI_T values of 8 or below indicated pollution, 12 or above, a healthy stream, and intermediate values, a semi-polluted state (Cairns and Dickson, 1971).

Generic Identification

Reexamine the slides from both sites under the oil immersion lens. Using a key (e.g., FWPCA, 1966), identify the more common genera and estimate their abundance.

QUESTIONS

(1) What are the dominant genera identified at the two stations?
(2) How do the two stations compare with regard to density, diversity, and composition?
(3) Plot genera distribution curves for both stations. How do these compare with those of Patrick (1962)?
(4) How do these density values compare with those generated in Exercise 3?

REFERENCES

Cairns, J., Jr. and K. L. Dickson. "A Simple Method for the Biological Assessment of the Effects of Waste Discharges on Aquatic Bottom-Dwelling Organisms," *J. Wat. Pollut. Cont. Fed.*, 43:755-772 (1971).

Cairns, J., Jr., D. W. Albaugh, F. Busey and M. D. Chanay. "The Sequential Comparison Index—A Simplified Method for Non-Biologists to Estimate Relative Differences in Biological Diversity in Stream Pollution Studies," *J. Wat. Pollut. Cont. Fed.*, 40:1607-1613 (1968).

Fjerdingstad, E. "Some Remarks on a New Saprobic System," *Biological Problems in Water Pollution, 3rd Seminar.* Public Health Service Publication No. 999-WP-25, pp. 232-235 (1962).

FWPCA. *A Guide to the Common Diatoms at Water Pollution Surveillance System Stations.* Federal Water Pollution Control Administration, Cincinnati, Ohio, 101 pp. (1966).

Kolkwitz, R. and M. Marsson. "Ökologie der Pflanzlichen Saprobien," *Ber. Deutschen Bot. Gesell*, 26a:505-519 (1908).

Patrick, R. "Algae as Indicators of Pollution," *Biological Problems in Water Pollution, 3rd Seminar.* Public Health Service Publication No. 999-WP-25, pp. 225-230 (1962).

Patrick, R., M. H. Hohn and J. H. Wallace. "A New Method for Determining the Pattern of the Diatom Flora," *Not. Nat. Acad. Nat. Sci. Philadelphia*, No. 259, 12pp. (1954).

Preston, F. W. "The Commonness, and Rarity, of Species," *Ecology*, 29:254-283 (1948).

EXERCISE 7/ALGAL CHLOROPHYLL QUANTITATION
BY ROBERT W. WALKER

PURPOSE

Algal biomass may be determined in several ways. Cells may be collected and weighed, the surface area or biovolume may be determined microscopically, or the concentration of various metabolic cofactors may be measured (ATP, chlorophyll *a*). In this exercise, two methods for the determination of chlorophyll from periphyton will be described: the acidification method and the trichromatic method.

INTRODUCTION

Since chlorophyll *a* comprises 1 to 2% of the dry weight of most algal cells and since it is easily quantified spectrophotometrically, this photosynthetic pigment is widely used for the determination of phytoplankton and periphyton biomass (APHA, 1985). Pheophytin *a*, a degradation product of chlorophyll *a*, has absorption peaks very similar to its parent compound. The fact that very dilute acid (0.003M HCl) is sufficient to remove the magnesium from the porphyrin ring of chlorophyll *a* and convert it to pheophytin *a* is utilized to determine the concentrations of both compounds (Lorenzen, 1967) and infer the physiological condition of the algal community (APHA, 1985). When using the acidification method, care must be taken to avoid adding excess acid as this would shift the absorption spectra of other plant pigments to those used for quantifying chlorophyll *a*. By measuring the absorption of the extract before and after acidification, the amount of chlorophyll *a* and pheophytin *a* may be calculated.

The second spectrophotometric method to be used in this exercise, the trichromatic method, is based upon the facts that the three major algal chlorophylls (*a, b,* and *c*) have different absorption maxima in the red portion of the spectrum and that pigment absorbancy is directly proportional to its concentration. This method, developed by Richards and Thompson (1952) and modified by Parsons and Strickland (1963), suffers from many sources of error and is not recommended for biomass or productivity measurements (APHA, 1985). It tends to overestimate the chlorophyll *a* content, particularly when no correction for pheophytin *a* is made and when the pigments are extracted from algal communities in poor physiological condition. In any event, this method, which is easily performed, will be used in conjunction with the acidification method to determine the chlorophyll *a* content of the periphyton colonizing the slides at our two sites.

PROCEDURE

Using the other half of the slides exposed in Exercise 6, determine the chlorophyll content of the colonized periphyton at each site. In the field, a razor blade is used to scrape the material from one surface of the slide into a tube containing 2 ml of 90% acetone [100% reagent grade acetone diluted 10% (v/v) with a saturated solution of $MgCO_3$ (1.0 g finely powdered/100 ml distilled water)]. Cover the tubes and store on ice in the dark until they can be returned to the laboratory for processing and analysis. Macerate the sample in a glass/glass or Teflon/glass tissue grinder. Transfer the sample to a centrifuge tube, rinse the tissue grinder adding the rinse to the sample, and bring the volume to either 5.0 or 10.0 ml using 90% acetone for all rinses and dilutions. Tightly seal the tube and steep the samples on ice in the dark for at least two hours (preferably overnight). Clarify by centrifuging in a closed tube for twenty minutes at 500 g.

To allow the determination of the pigment concentrations by both methods on a single sample, the following procedure should be used. Transfer 3.0 ml

Algal Chlorophyll Quantitation

of the extract to a cuvette with a 1-cm path length and read the optical density (OD) at 750, 664, 647, and 630 nm. Record these values in Table 7-1. Acidify the extract in the cuvette by adding 0.1 ml of 0.1 N HCl and mix the contents gently. Ninety seconds after acidification, read the OD at 665 and 750 nm and record the values in Table 7-1.

To remove the effect of any turbidity in the sample, subtract the appropriate 750 nm OD from the respective OD values at 664, 647, 630, and 665 nm. Record the adjusted optical densities in Table 7-1 and use these values in the following calculations.

TABLE 7-1 Extract optical densities.

Sample	OD Before Acidification				OD After Acidification		Optical Densities Adjusted for Turbidity			
	750	664	647	630	665	750	OD664	OD647	OD630	OD665
Control 1										
2										
3										
4										
5										
6										
7										
8										
9										
10										
Test 1										
2										
3										
4										
5										
6										
7										
8										
9										
10										

CALCULATIONS

For the following calculations to be valid, the optical densities must be determined with the sample in a cuvette with a 1 cm path length.

Acidification Method

Determine the chlorophyll *a* and pheophytin *a* concentrations in mg pigment/m² of substrate surface area by:

$$\text{chlorophyll } a = \frac{26.7 \times (\text{OD664} - \text{OD665}) \times V}{A}$$

$$\text{pheophytin } a = \frac{26.7 \times [(1.7 \times \text{OD665}) - \text{OD664}] \times V}{A}$$

where,

OD664 = the adjusted OD of the extract at 664 nm
OD665 = the adjusted OD of the acidified extract at 665 nm
   V = the extract volume (l)
   A = the surface area of the substrate (m²). For a single surface of a 25 by 75 mm slide, the area is 0.001875 m²

Record the results in the appropriate spaces on Table 7-2.

Trichromatic Method

Determine the chlorophyll concentrations in mg pigment/l of extract by:

$$\text{chlorophyll } a = 11.85(\text{OD664}) - 1.54(\text{OD647}) - 0.08(\text{OD630})$$

$$\text{chlorophyll } b = 21.03(\text{OD647}) - 5.43(\text{OD664}) - 2.66(\text{OD630})$$

$$\text{chlorophyll } c = 24.52(\text{OD630}) - 7.60(\text{OD647}) - 1.67(\text{OD664})$$

where,

OD664 = the adjusted OD of the extract at 664 nm
OD647 = the adjusted OD of the extract at 647 nm
OD630 = the adjusted OD of the extract at 630 nm

TABLE 7-2 Pigment concentrations in mg/m².

Sample	Acidification Method		Trichromatic Method		
	chlor. *a*	pheo. *a*	chlor. *a*	chlor. *b*	chlor. *c*
Control 1					
2					
3					
4					
5					
6					
7					
8					
9					
10					
Test 1					
2					
3					
4					
5					
6					
7					
8					
9					
10					

The amount of pigment per unit surface area (mg pigment/m²) is calculated by:

$$\text{chlorophyll } a, b, \text{ or } c = \frac{C \times V}{A}$$

where,

C = the concentration of the respective pigment in mg/l of extract (from above)

V = the extract volume (l)
A = the surface area of the substrate (m²)

Record the results in the appropriate space on Table 7-2.

QUESTIONS

(1) Do you observe any differences in the pigment concentrations between the two sites? Speculate.

(2) Compare the chlorophyll *a* concentrations, as determined by the acidification and the trichromatic methods, for your data from the control site. Do differences exist between the methods? Do differences exist for the test site? Explain.

REFERENCES

APHA. *Standard Methods for the Examination of Water and Wastewater*, 16th ed. American Public Health Association, Washington, D.C., 1268pp (1985).

Lorenzen, C. J. "Determination of Chlorophyll and Pheo-Pigments: Spectrophotometric Equations," *Limnol. Oceanogr.*, 12:343–346 (1967).

Parsons, T. R. and J. D. H. Strickland. "Discussion of Spectrophotometric Determination of Marine-Plant Pigments, with Revised Equations for Ascertaining Chlorophylls and Carotenoids," *J. Mar. Res.*, 21:155–163 (1963).

Richards, F. A. and T. G. Thompson. "The Estimation and Characterization of Plankton Populations by Pigment Analyses. II. A Spectrophotometric Method for the Estimation of Plankton Pigments," *J. Mar. Res.*, 11:156–172 (1952).

CHAPTER 3

Community Function

INTRODUCTION

THE application of community function, either independently or as an adjunct to community structure, provides an ecological dimension to assess the response to stress. Because altered performance is a gradient rather than a quantum measure, it offers a more sensitive assay, when digested statistically, than the mere presence or absence of an organism. Furthermore, by obviating the need for identification, the investigator can generate data rapidly and inexpensively. As such, it offers pollution control agencies a viable alternative to supporting expensive taxonomic expertise. Among the life functions you will investigate in these exercises are photosynthesis and respiration.

In a sense, this approach, as a transition between the laboratory and the field, is a healthy compromise. It provides greater control than field surveys and more applicability than laboratory research.

EXERCISE 8/PRIMARY PRODUCTIVITY AND PHOTOSYNTHETIC EFFICIENCY USING THE LIGHT-DARK BOTTLE METHOD

PURPOSE

This exercise will introduce you to the light-dark bottle method of measuring primary productivity by studying a simulated lake in the laboratory (Exercise 1). Also, a technique used for determining the photosynthetic efficiency of the phytoplankton in the lake will be presented.

INTRODUCTION

The fundamental function that fuels the negative entropic processes of the biosphere is primary productivity: the creation by photosynthesis of organic matter through the reduction of inorganic carbon (Lieth, 1975). The organic matter created can be expressed as standing crop: the biomass per unit area at any point in time (Wetzel, 1983). The higher the community's productivity, the greater the amount that is likely to end up as biomass (Lieth, 1975). Consequently, biomass determinations have been applied as estimates for determining the trophic state of lakes. Such data, however, are of limited value when applied to species with short generation times, since this rate of synthesis is not reflected in increased mass. Productivity measurements, therefore, are more valid indices of the trophic state of lakes. Wetzel (1983) has used productivity rates for this purpose and has compiled a table of the general ranges for the approximate net primary productivity of phytoplankton associated with the various trophic states. These rates range from less than 50 mg C/m²·day for ultraoligotrophic, 50 to 300 mg for oligotrophic, 250 to 1000 mg for mesotrophic to greater than 1000 mg for eutrophic lakes.

The use of phytoplankton primary productivity exclusively as a criterion for trophic state ignores autochthonous inputs from the littoral zone (periphyton and macrophytes) and allochthonous detritus sources. Nevertheless, the primary productivity of phytoplankton remains a significant factor in the assessment of lake metabolism (Wetzel, 1983).

The energy derived from photo-reduction is used to assimilate nitrogen, phosphorous, and sulfur-containing compounds. This process has been summarized by Odum (1971) as: 1.3×10^6 kcal of radiant energy + 106 CO_2 + 90 H_2O + 16 NO_3 + 1 PO_4 + mineral elements = 1.3×10^4 kcal of potential energy represented by 3258 g of protoplasm (106 C, 180 H, 46 O, 16 N, 1 P, and 815 g of mineral ash) + 154 O_2 + 1.287×10^6 kcal or 99% dispersed as heat. Thus, only 1 percent of the sun's energy falling on a terrestrial plant is fixed as potential energy. In aquatic systems the efficiency is usually an order of magnitude less. A portion of the sunlight is reflected at the water's surface and of that which penetrates, a significant portion is absorbed by particulate matter and dissipated as heat.

The essential relationship governing primary production can be expressed as:

$$6\ CO_2 + 6\ H_2O \xrightarrow[\text{light receptor}]{\text{light}} C_6H_{12}O_6 + 6\ O_2$$

Therefore, we can assume that for every molecule of oxygen released, one atom of carbon is fixed or assimilated into organic matter. The above equation can be written more generally as:

$$CO_2 + 2\,AH_2 \xrightarrow{h \cdot v} HCOH + 2\,A + H_2O$$

where, AH_2 represents a H_2 donor which can be H_2S, H_2O, or a reduced organic compound (Vollenweider, 1974). The latter equation encompasses all photoautotrophs, rather than just the aerobic contributors. These equations lead to several defensible methodological approaches for determining primary productivity rates. Measurement of oxygen production, carbon uptake, the formation of organic material, the increase in chemical energy of the system, or even the changes in the cell redox system are all theoretical possibilities.

The most commonly used method, the light-dark bottle technique, is predicated on determining dissolved oxygen levels in clear and opaque BOD bottles containing lake water samples suspended and incubated at selected depths. The variation in productivity over depth can be used to determine a depth profile, the integration of which yields the total productivity (mg C/m²·day) of a column of water.

Although the light-dark bottle technique yields a reasonable estimate of phytoplanktonic production, there are some inherent problems that must be kept in mind. These shortcomings result from the need to incorporate generalizations that are not entirely accurate. This approach assumes that: (1) Respiration rates are the same in the light and dark, and that bacterial and animal respiration demands are small in relation to those of the planktonic algae (Wetzel and Likens, 1979; Darley, 1982). (2) No other processes, including photo-oxidative consumption, utilize oxygen (Vollenweider, 1974). (3) Such environmental variables as light and temperature are the same inside and outside the bottle. However, because the adsorption of light by glass, especially in the shorter wavelengths, is different than that of the surrounding water, the "light climate" inside the bottle is different (Vollenweider, 1974). Additionally, there are the acknowledged differences in the internal isolated conditions (i.e., turbulence, nutrient levels, etc.). These, however, can be mitigated by reducing the incubation period from the usual 24 hours to 4 to 6 hours. (4) The phytoplankton are responsible for the total productivity of the lake. This is not true, because in many cases, the littoral production of macrophytes and periphyton, as well as that of allochthonous material can be large in relation to phytoplanktonic productivity (Odum, 1971; Wetzel, 1983). These facts notwithstanding, the technique has been widely applied for estimating primary productivity and if kept in proper perspective, it can be an exceedingly useful tool.

Little research has been reported relating photosynthetic efficiency, a measure of a plant's capacity to convert solar energy into chemical energy (Jones, 1979), to water quality. Any change in water quality which alters the amount of light energy entering the system will affect productivity, and any change which affects productivity without changing the light parameter will affect efficiency. Frequent monitoring of the efficiency can, therefore, detect a change in water quality due to external interferences.

PROCEDURE

Bottles

It is important that the bottles are thoroughly clean and that all traces of acid and nutrients are removed prior to use. Dark bottles can be prepared by painting them black or by covering them with two layers of black electrical tape. Once filled, the stopper and neck should be covered with aluminum foil to exclude all light.

Sampling and Incubation

For each depth, a set of one dark and two clear BOD bottles will be used. Without greatly disturbing the water column, measure the depth and temperature from the levels where the samples will be taken and record them in Table 8-1. Insert the siphon tube, from the appropriate sample depth (epilimnion, metalimnion, or hypolimnion), to the bottom of the bottle and allow three volumes of water to overflow. Carefully remove the tube and immediately stopper the bottle remembering to cover the stopper of the dark bottle with aluminum foil. Immediately fix one of the clear bottles from each depth using the reagents for the Alsterberg azide modification of the Winkler method (APHA, 1985). Suspend the remaining bottles at the appropriate depths in the "lake" and incubate them for 2 to 6 hours depending upon the projected productivity level.

Determine the initial dissolved oxygen (DO) levels (I) and record them in Table 8-1. After incubation, determine the DO concentrations for the remaining light (L) and dark (D) bottles and record these values in Table 8-1.

Light Measurement

Adjust the height of the light source to minimize the heat effect and to maximize the evenness of illumination over the water's surface. Using a photometer, take readings at points on the surface in a grid fashion and record

TABLE 8-1 Raw data sheet.

Level	Depth (m)	Temperature (°C)	Incubation Time (hours)	Dissolved Oxygen (mg/l) Initial	Light	Dark
Epilimnion						
Metalimnion						
Hypolimnion						

them. The light source used should give photometer readings of approximately 200 ft-candles (2152 lux). Light intensity contours are drawn between the measured values and an average value is derived.

The percent of incident light with depth can be determined by dividing the mean values at the selected depths by the mean surface value and multiplying by 100. Prepare a graph plotting the percent of incident light at the preselected depths versus depth.

CALCULATIONS

Primary Productivity

Determine photosynthesis and respiration in mg O_2/l·hour by:

$$\text{net photosynthesis} = \frac{L - I}{t}$$

$$\text{respiration} = \frac{I - D}{t}$$

$$\text{gross photosynthesis} = \frac{L - D}{t}$$

where,

I = the initial DO in mg O_2/l
L = the DO of the incubated light bottle in mg O_2/l
D = the DO of the incubated dark bottle in mg O_2/l
t = the incubation period in hours

Record the data in Table 8-2.

TABLE 8-2 **Productivity and efficiency.**

Level	Photosynthesis/Respiration (mg O_2/l·hour)			Gross Primary Productivity (mg C/m^3·day)
	Net	Gross	Respiration	
Epilimnion				
Metalimnion				
Hypolimnion				
Lake Gross Productivity (g C/m^2·day):				
Photosynthetic Efficiency (%):				

The gross primary productivity in mg C/m^3·day is calculated as:

$$\text{gross primary productivity} = \frac{L - D}{t} \times \frac{12}{32} \times 1000 \times 12$$

where,

L = the DO of the incubated light bottle in mg O_2/l
D = the DO of the incubated dark bottle in mg O_2/l
t = the incubation period in hours
$\frac{12}{32}$ = the atomic weight of carbon/molecular weight of oxygen
1000 = the conversion factor for liters to cubic meters
12 = the hypothetical number of hours of light per day

Record the results in Table 8-2.

Plot the gross primary productivity (mg C/m^3·day) versus depth (m). The area within the curve is determined to estimate the gross primary productivity of the euphotic zone in mg C/m^2·day. Divide this value by 1000 and record the result, in g C/m^2·day, in Table 8-2.

Photosynthetic Efficiency

Measuring light energy in units of irradiance (energy flux per unit area and time, e.g., cal/cm²·sec, ergs/cm²·sec, etc.) with a pyrheliometer is preferable to determining the illuminance (light intensity, e.g., lux or ft-candles) with a photometer because of the latter instrument's limited spectral response and the difficulty of direct conversion between units. Since in this exercise a photometer was used to measure illuminance, it will be necessary to convert the value obtained for light intensity (ft-candles) to energy units of photosynthetically active radiation (cal/m²·day). This is accomplished by multiplying the photometer reading by 889 for natural daylight, 622 for fluorescent lamp light, or 1200 for tungsten lamp light. These conversion factors are approximations based upon those cited by Vollenweider (1974).

The values obtained for primary productivity and for average light energy are used to determine the phytoplanktonic photosynthetic efficiency using a modification of the equation of Lind (1979).

$$\text{photosynthetic efficiency (\%)} = \frac{P \times 2 \times 5500}{LE} \times 100$$

where,

P = the primary productivity in g C/m²·day
2 = the approximate conversion of carbon to dry algal tissue
5500 = approximate caloric equivalent per gram dry algal tissue
LE = the light energy of the photosynthetically active portion of the spectrum in cal/m²·day. The 0.5 usually found in the denominator of this formula can be ignored because the above conversion factors determine the light energy that is present in the photosynthetically active portion of the spectrum.

Record the result in Table 8-2.

For freshwater systems, the photosynthetic efficiency is usually less than 1 to 2% (Lind, 1979). Values as high as 4% have been observed in eutrophic lakes (Wetzel, 1983).

QUESTIONS

(1) If these were data from the topics, would you consider the lake eutrophic? Explain.
(2) How could this technique be adapted to the field (i.e., how would you go about setting up this experiment to determine the productivity of a lake)? How would you determine the productivity for an entire day? A month? A year? How would you propose to sample from each depth? What factors would you consider important (i.e., probable sources of error)?
(3) What productivity rates would you expect in polluted rivers?
(4) How do you account for the differences in productivity among the strata?

(5) Does the tank exhibit surface photo-inhibition, biogenic turbidity, or any other distinct profile anomolies?
(6) Identify, if possible, the compensation level, trophogenic and tropholytic zones.

REFERENCES

APHA. *Standard Methods for the Examination of Water and Wastewater*, 16th ed. American Public Health Association, Washington, D.C., 1268 pp. (1985).

Darley, W. M. *Algal Biology: A Physiological Approach*. Blackwell Scientific Publications, Oxford, 168 pp. (1982).

Jones, G. *Vegetation Productivity*. Longman Publishing, London, 100 pp. (1979).

Lieth, H. "Primary Production of the Major Vegetation Units of the World," in *Primary Productivity of the Biosphere*, H. Lieth and R. H. Whittaker, eds. Springer-Verlag, New York, pp. 203–215 (1975).

Lind, O. T. *Handbook of Common Methods in Limnology*, 2nd ed. C. V. Mosby Co., St. Louis, 199 pp. (1979).

Odum, E. P. *Fundamentals of Ecology*, 3rd ed. W. B. Saunders Co., Philadelphia, 574 pp. (1971).

Vollenweider, R. A. *A Manual on Methods for Measuring Primary Production in Aquatic Environments*, 2nd ed. Blackwell Scientific Publications, Oxford, 225 pp. (1974).

Wetzel, R. G. *Limnology*, 2nd ed. Saunders College Publishing, Philadelphia, 767 pp. (1983).

Wetzel, R. G. and G. E. Likens. *Limnological Analyses*. W. B. Saunders Co., Philadelphia, 357 pp. (1979).

EXERCISE 9/A FIELD METHOD FOR ASSESSING THE WATER QUALITY OF A STREAM USING PRIMARY PRODUCTIVITY

PURPOSE

The purpose of this exercise is to demonstrate a procedure for determining the primary productivity of periphyton colonizing tubular substrates in a stream. While this method does not support extrapolation to the stream bed, it does serve as a monitoring tool allowing for the routine diagnosis of water quality.

INTRODUCTION

Primary productivity may be defined as the rate at which radiant energy is utilized by the photosynthetic activity of producers to form organic substances of high potential energy from inorganic sources (CO_2 and H_2O). Productivity is a measure of the ability of a system to replace or increase its organic content. In simple terms, it is a measure of the rate of growth. Biomass or standing crop—the amount of organic matter supported by a system at any one moment—although often used for this purpose, is a poor index of productivity without some indication of the time involved in its formation (Ryther, 1956).

The community of primary producers is different with each type of aquatic system. Phytoplankton are considered to be the dominant primary producers in lacustrine systems (Cole, 1983), although epiphytic algae and macrophytes can be an important source of organic matter in lakes with a large littoral zone (Wetzel, 1983). Periphyton are the principle autochthonous primary producers in streams (Hynes, 1970; Smith, 1980), where macrophytes are generally unable to establish themselves in the unstable bed and inconsistent flow, and phytoplankton, due to their nonsessile nature, are carried with the current and require lengthy reaches to reproduce. Macrophyte and phytoplankton productivity become more important with increasing stream size.

In lotic systems the significance of the benthic primary producers varies (Hansmann et al., 1971). In streams where food chains are primarily grazing rather than detritus based, the principal primary producers are periphytic algae. As allochthonous detritus and turbidity increase, the contribution of this community decreases. The activity and abundance of this base of the energy pyramid is an index, therefore, of the biological status of an aquatic system. Production of the periphyton community closely reflects the water quality (APHA, 1985), immediately responding to a confluence, diffuse runoff, or an outfall (Tilley and Haushild, 1975).

Consequences of productivity are most easily seen in eutrophic environments. The primary producers are able to utilize nutrient and organic compounds not only from the aquatic system, but also from watershed runoff. If this runoff contains high levels of nutrients, it will markedly increase productivity. This enrichment can grossly alter the populations of aquatic organisms within the system by selecting for r-strategists which favor increased productivity (Smith, 1980).

While photosynthesis is occurring, there is oxygen uptake due to respiration which is further compounded by oxygen exchanges with the atmosphere, groundwater, and surface runoff (Odum, 1956; Hynes, 1970). For these reasons, it is particularly difficult to extrapolate to productivity rates in flowing water from dissolved oxygen levels. How does the investigator weigh each factor quantitatively? Sampling to measure oxygen and temperature at one to two-hour intervals for a 24-hour period attempts to address this dilemma (Odum, 1956; APHA, 1985). While the logic is irrefutable, the protocol is time-consuming, expensive, and subject to error because of the estimates required for accrual and diffusion rates.

Through the use of tubes as artificial substrates, however, accrual and diffusion rates can be ignored and the flow regime precisely measured. The investigator only needs to measure the flow rate and the oxygen levels at both ends of the tube to estimate net productivity. Further, by covering a colonized tube with an opaque tube, a diurnal regime can be simulated – the rationale being the same as for the dark bottle: to determine respiration rates and from this gross productivity.

This method has been used to assess the impact of nutrient enrichment (Coler and Swift, 1980; Tease et al., 1983) and coal leachate (Tease and Coler, 1984) on periphyton productivity. To use this method, one set of colonized tubes is placed upstream of the pollution source (control) and another set downstream (test). The dissolved oxygen (DO) is measured in each set and productivity calculated. The current inside the two tube sets must be comparable, but need not equal the stream flow because in this case, we are concerned only with relative differences. This procedure readily affords comparisons between substrates at different locations in the stream; however, it does not support extrapolation to the stream bed (Tease et al., 1983).

PROCEDURE

Installation and Colonization

At control and test site stations with similar environmental conditions (depth, flow, temperature, light, etc.), assemble the substrate supports. These consist of three horizontal wooden poles lashed to stakes driven into the stream bed which support the tubes in the middle and at each end. The tubular substrates consist of a 2.4 m section of 0.9 cm inside diameter polycarbonate tubing with a 15 cm diameter plastic funnel covered with wire screening attached (removable) to the upstream end of the tube. Label each tube in a manner that will allow its identity to be known throughout both the baseline and experimental portions of the research. Using string or wire, secure all six substrates to the support at the control site with the downstream end held roughly 4 to 6 cm higher than the upstream end to prevent blockage by air bubbles. Allow the tubes to colonize for two to three weeks.

Baseline Productivity

Following colonization, determine the relative baseline productivity (net production of oxygen) for each tube a minimum of five times over five or more days. Record the appropriate data in Table 9-1.

TABLE 9-1 **Baseline productivity.**

Date Hour Temp	Tube Code	Carboy DO 1	2	$\bar{x}$	Outflow DO 1	2	$\bar{x}$	DO Differential (mg O_2/l)	Flow (l/min)	Net Productivity (mg O_2/min)

Prior to collecting the samples, the tube should be wiped clean of external growth, and the funnel assembly should be removed. A 20 l carboy, filled to a designated level and fed via a battery-powered water pump [or a Mariotte bottle (Burrows, 1949) made from a 20 l carboy], is connected to the upstream end of each substrate to provide stream water at the appropriate flow rate. The downstream end of the substrate is supported 15 to 20 cm above the stream surface and a 20 cm section of tubing is attached. Using a graduated cylinder and a stopwatch, adjust the flow rate to 5 or 6 ml/second by using either the carboy spigot or changing the outflow height and record the flow rate (l/minute) for each tube in the appropriate table. While maintaining this rate of flow, collect duplicate water samples for each tube in 300 ml BOD bottles by inserting the tubing attached to the downstream end of the substrate to the bottom of the bottle. Allow the bottle to overflow for roughly two minutes, then slowly remove the tubing from the bottle, and stopper it carefully to avoid trapping air bubbles. Similarly, collect duplicate samples from the carboy. Each sample should be fixed immediately (in the field) and titrated upon return to the laboratory using the Alsterberg azide modification of the Winkler method (APHA, 1985) to determine the DO concentration. Record the carboy and outflow DO values in the appropriate table and determine the mean ($\bar{x}$) for each. Net productivity for the tube (mg O_2/minute) is determined by multiplying the DO concentration differential ($\bar{x}$ outflow $-\bar{x}$ carboy) in mg O_2/l by the flow rate in l/minute.

Experimental Productivity

After a minimum of five days of monitoring the baseline productivity of all six tubes at the control site, stopper both ends of three of the tubes (selected at random), transfer them to the test site, and secure them to the previously constructed support. To assure that all tubes are equally stressed by the moving process, stopper the remaining three tubes, carry them a similar distance, and resecure them to the support at the control site. Then, using the previously described procedure, measure the relative net productivity for each tube a minimum of five times over five or more days. Record the appropriate data in Table 9-2.

Calculate the mean net productivity with confidence limits for each set of three tubes for each time interval during both the baseline and experimental periods. Construct a graph, plotting time versus mean net productivity for both treatments.

TABLE 9-2 **Experimental productivity.**

Date Hour Temp	Tube Code	Carboy DO 1	2	$\bar{x}$	Outflow DO 1	2	$\bar{x}$	DO Differential (mg O_2/l)	Flow (l/min)	Net Productivity (mg O_2/min)

Community Structure

Cut a 5 cm section from comparable sections of all the tubes at the start, middle, and close of the experimental period. Using the methods outlined in Exercise 6, prepare slides and determine the dominant genera present in the samples.

QUESTIONS

(1) Are there different dominant genera in the two substrate sets?
(2) Are the stations different with regard to net productivity?
(3) How would you determine gross productivity?

REFERENCES

APHA. *Standard Methods for the Examination of Water and Wastewater*, 16th ed. American Public Health Association, Washington, D.C., 1268 pp. (1985).

Burrows, R. E. "Prophylactic Treatment for Control of Fungus *Saprolegnia parasitica* on Salmon Eggs," *Prog. Fish Cult.* 11:97–103 (1949).

Cole, G. A. *Textbook of Limnology*, 3rd ed. C. V. Mosby Co., St. Louis, 401 pp. (1983).

Coler, R. A. and L. Swift. "Demonstrating the Limiting Nutrient Content in Rural River Reaches," *Am. Biol. Teach.* 42:353–355 (1980).

Hansmann, E. W., C. B. Lane and J. D. Hall. "A Direct Method of Measuring Benthic Primary Production in Streams," *Limnol. Oceanogr.* 16:822–826 (1971).

Hynes, H. B. N. *The Ecology of Running Waters*. Liverpool University Press, Liverpool, 555 pp. (1970).

Odum, H. T. "Primary Production in Flowing Waters," *Limnol. Oceanogr.* 1:102–117 (1956).

Ryther, J. H. "The Measurement of Primary Production," *Limnol. Oceanogr.* 1:72–84 (1956).

Smith, R. L. *Ecology and Field Biology*, 3rd ed. Harper & Row Publishers, New York, 835 pp. (1980).

Tease, B. and R. A. Coler. "The Effect of Mineral Acids and Aluminum from Coal Leachate on Substrate Periphyton Composition and Productivity," *J. Freshw. Ecol.* 2:459–467 (1984).

Tease, B., E. Hartman and R. A. Coler. "An *In Situ* Method to Compare the Potential for Periphyton Productivity of Lotic Habitats," *Water Res.* 17:589–591 (1983).

Tilley, L. J. and W. L. Haushild. "Use of Productivity of Periphyton to Estimate Water Quality," *J. Wat. Pollut. Cont. Fed.* 47:2157–2171 (1975).

Wetzel, R. G. *Limnology*, 2nd ed. Saunders College Publishing, Philadelphia, 767 pp. (1983).

CHAPTER 4

Bioassay and Toxicity Testing

AQUATIC TOXICOLOGY: PRINCIPLES AND PROCEDURES[1]

PURPOSE

THIS unit will introduce you to those procedures practiced to evaluate the toxicity of chemical substances to aquatic life. Such toxicity data are presently used to determine permissible effluent discharge rates into the aquatic environment and to monitor levels of contaminants in streams with respect to water quality standards.

INTRODUCTION

The terms "bioassay" and "toxicity test" have frequently, but incorrectly, been used interchangeably. They, in fact, have two distinct meanings (Brown, 1973; Stephan, 1977). Toxicity tests determine the concentration of a chemical which causes either death or some altered physiological process reflecting interference with the normal life cycle of the test organism. The objective of a bioassay, on the other hand, is to adapt the biological response to determine the concentration of a specific chemical entity. It is, therefore, essentially the same as a chemical assay or analytical chemical measurement. Typically, bioassays are applied to determine the amount of pesticide, metal, vitamin, or other substance in a sample. The results of the bioassay are then converted to concentration values based on a standard curve.

Alternatively, toxicity data provide information about the mechanisms of toxicity, synergistic or antagonistic interactions with various environmental

[1]Taken, in part, from Peltier, 1978; Medeiros et al., 1981; Plotkin and Ram, 1983.

parameters, and the overall impact of stresses with respect to mortality, growth, reproduction, respiration, excretion, and behavior. Consequently, the data are used to determine the level of waste treatment required to meet pollution control regulations and to evaluate the effectiveness of waste treatment methods.

Toxicity tests have been performed by numerous investigators for well over a hundred years. Unfortunately, most of the data collected prior to 1951 are virtually useless due to the lack of a standardized technique that allows comparison. Doudoroff et al. (1951) evolved a protocol that details the procedure to be followed and emphasizes knowledge of the test water quality.

GENERAL METHODS AND PROCEDURES OF TOXICITY TESTING

Brown (1973) recognized three classes of pollutant materials that may be examined using toxicity tests: (1) materials essentially chemical in nature such as metal salts and hydrogen cyanide, (2) physical materials including mineral particulates, radionuclides, hot water, and hypertonic solutions, and (3) biological materials such as viruses and bacteria.

In our studies we will focus on the first class of contaminants which constitutes most of the pollution problems found in the aquatic environment. This is attributable to the large amount of synthetic residue introduced into the environment and the inability of most aquatic organisms to survive or continue their normal life functions in the presence of these substances.

Toxicity tests are conducted by exposing fish, invertebrates, or algal species to varying concentrations of the toxicant for some period of time. Upon termination of the test, the percent mortality is observed at each toxicant concentration, and the LC50 (median lethal concentration) is calculated. The percent mortality at varying exposure times may be expressed using the LC symbol. For example, a 96-hour LC50 is the concentration causing 50% mortality in 96 hours. The use of LC followed by a percentage number allows the designation of percentages of mortality other than 50%. For example, one can determine a LC90 to remove undesirable species from a fishing area, or a LC10 to insure survival of fish exposed to industrial wastes. However, the LC50 is the standard measure of toxicity because, for a fixed number of experimental units, the median response tends to be the most reliable (Finney, 1971). (The LD50, used in pharmacology and mammalian toxicology, differs from the LC50 in that the toxicant dosage is introduced directly into the test organism.) The converse of the LC50 term is the TLm (median tolerance limit) which is the concentration, after some period of exposure, where 50% survival is observed. An additional expression is the EC50 or median effective concentration which is used to designate a level that depresses some life function or process (i.e., reproduction, growth, respiration, etc.) by a factor of 50%.

Flow Regimes

Toxicity tests can be divided into three groups with respect to flow: static, continuous flow, and renewal. All three types may be modified by the inclusion of a circulatory pump to simulate flowing water and stimulate animal activity.

Static Toxicity Tests

These tests are simple and easily controlled. They are suitable for the detection and evaluation of compounds whose toxicity is associated with low oxygen demand and high stability. The major limitations with static tests are that the toxic substance being tested may be removed by volatilization, precipitation, or detoxification by the test organisms and that the test must be relatively short-term because of the stress caused by the accumulation of metabolic byproducts in a totally alien environment.

Flow-Through Tests

These tests are more complicated in design, but new developments in apparatus and technique have made their usage practical. They are mainly employed to test chronic toxicity of industrial effluents and chemicals that have high biochemical oxygen demands, are unstable or volatile, or are appreciably removed from solution by precipitation or by adsorption to the test organism or vessel. The protocol provides for well-oxygenated solutions, nonfluctuating concentrations of toxicant, and removal of metabolic byproducts. Furthermore, it more closely approximates the natural conditions of receiving streams and permits extended exposures to detect chronic toxicity and identify "safe" concentrations of the toxicant. The shortcomings inherent in the flow-through approach are the large amount of space needed, the complicated experimental apparatus, and the large amounts of wastes and diluent waters that are required.

The first two of these limitations were alleviated by the development of the simple and inexpensive serial dilution apparatus pioneered by Mount and Brungs (1967). In such an apparatus, incoming water is mixed with the toxicant at a specified proportion by a diluter. This method assures a constant concentration of toxicant and oxygen in the test water, regardless of toxicant volatility or oxygen demand.

Unlike the static tests, the protocol for flow-through testing permits feeding without the danger of fouling or toxicant interaction. This allows testing of the organisms for their entire life cycle to determine mortality as well as productivity, mutagenicity, and teratogenicity.

Renewal Toxicity Tests

This methodology can be considered as intermediate between static and flow-through. It differs from the former by periodically renewing the test solution. The test concentration is maintained by abrupt and perhaps stressful, periodic renewals of the test water, not gradually as in the continuous flow test.

Changing the test solution for the purpose of keeping more or less uniform concentrations of any volatile or unstable toxic materials and an adequate dissolved oxygen level can be accomplished by quickly transferring the test organisms to fresh solutions. Here, as in the flow-through test, the organisms can be fed without the accumulation of metabolites and excess food. Test solutions should be renewed every twenty-four hours or less.

Selection of Species

Test animals must be adaptable to laboratory conditions and available in adequate numbers of the appropriate size. When possible, the test animal should be native to the watershed under investigation, and preferably, a species of recreational or commercial value. If these conditions cannot be met, the investigator may be compelled to obtain the test individuals from a commercial dealer. In such instances, comparative tests should be performed to relate the sensitivity of the selected test organism to the most sensitive endemic species. It is important that any unusual conditions to which the test group was exposed (e.g., pesticides or chemotherapeutic agents) be considered and reported (APHA, 1985).

APHA (1985) lists additional factors that should be considered in choosing a test organism including: (1) their sensitivity to the materials or environmental conditions under consideration, (2) their geographical distribution, abundance, and availability throughout the year, (3) the availability of culture methods and knowledge of their environmental requirements, (4) their general physical condition and freedom from parasites and disease, (5) the amount of space and time required because of the organism's size and the length of its life cycle, and (6) their amenability to comparison with other toxicity data obtained in previous research or toxicity investigations.

Organism Size

The length of the test organisms should not vary in extremes by more than 50% between the smallest and the largest. For fish, however, researchers contend that a variation of more than a few millimeters is too great because the weight of a fish increases as the cube of its length. A natural population of fish, however, is varied in size, and the application of toxicity data derived from a very uniform group of fish to problems involving natural populations may be questionable. The test animals must also belong to the same age class as well as size class. Immature specimens are commonly used because differences attributable to sex are minimized.

Preparation and Acclimation of Test Fish

Specimens selected for testing should be acclimated to the test temperature and the dilution water for at least ten days, but preferably thirty days. LC50 values for pH, DO, NH_3, and a number of noncumulative toxicants increase severalfold upon long-term (greater than five days) exposure to low levels or progressively increasing exposure to the test toxicant. Acclimation often occurs only below a certain concentration and can be transferable between similar toxicants. Effects upon chronic toxicity are virtually unknown. Imagine the consequence of stocking thousands of fish in a water body where a resident population has already acclimated to an existing level of pollution. Could the absence of extended acclimation contribute to the exceedingly low return rate of stocked trout in slightly polluted waters? Pre-exposure of fish to certain condition changes is a poorly studied area.

Acclimation to temperature has been investigated; however, different researchers have produced different results. The commonly used rule of thumb of "at least one day per centigrade degree of change" has no particular scientific basis and may sometimes contradict the facts. Peterson and Anderson (1969) concluded that the acclimation of metabolism to temperature change takes at least two weeks, regardless of direction. Presently, it seems the longer the period the better; however, variations in other conditions associated with temperature change in the laboratory may make this approach impractical.

General Test Conditions and Protocol

Temperature

The temperature range for test solutions must not exceed 4°C and must be appropriate; that is, 25°C for warm water fish and 15°C for cold water fish. Holding and test temperatures (±2°C) of 7, 12, 17, 22, and 27°C are recommended for fish and invertebrates by Peltier (1978). The organisms should be held and tested within 5°C of the temperature of the water from which they were obtained.

Dissolved Oxygen and Aeration of Test Solutions

The dissolved oxygen concentration must not fall below 5 mg/l and 4 mg/l for cold and warm water fish, respectively. Aeration with finely dispersed air is not permissible when it will cause reduction in toxicity of the solutions by accelerating the loss of volatile components. To avoid depletion of the dissolved oxygen, the weight of the animals in the container should not exceed one gram per two liters of test solution.

Concentrations of Toxicants Used

Concentrations of dilutions of liquid industrial wastes are expressed as percent by volume, while concentrations of nonaqueous wastes and of individual chemicals are presented in terms of milligrams per liter (mg/l) or micrograms per liter (μg/l). For testing, the concentrations are conveniently expressed on a logarithmic scale.

The inclusion of any waters of hydration as part of the weight of the solute, i.e., $CuSO_4 \cdot 5H_2O$, should be clearly indicated. When an impure chemical is used, all ingredients should be reported. The range of toxicant concentrations to be examined by the full-scale test should include at least two replicates of five to six concentrations. This range must first be determined by exploratory (screening) tests.

Controls

A control of diluent water (preferably ground or synthetic water) without any toxicant must be employed as a point for comparison. There should be no more than 10% mortality among the control animals during the course of any valid test with the remaining 90% appearing apparently vigorous. When an organic solvent or other dispersing agent is used to prepare test solutions, the test control should contain the maximum concentration of the solvent or dispersant to which the organisms in the other solutions are exposed.

Sample Size

At least 20 (ten per replicate) test organisms should be allotted to each experimental concentration. It is not essential to maintain two or more chambers per concentration during the test as the results are usually combined for analysis; however, this practice is advisable because: (1) viewing and counting are made easier, (2) overloading is less likely to occur, and (3) accidental invalidation of the test because of a broken test container is more easily avoided (Peltier and Weber, 1985).

Feeding

The daily feeding regimen should be interrupted two days before the initiation of a static test. Specimens should not be fed during tests of limited duration (ninety-six hours or less) to avoid large fluctuations in their metabolic rates and the fouling of the test solution with metabolic wastes and uneaten food. During a test of longer duration, however, food is required.

Test Duration and Observations

The duration of all tests should be at least forty-eight hours (96-hour tests are preferable). When more than one-half of the test fish survive at the highest concentration, the test must be continued to ninety-six hours.

The number of test fish in each container must be observed and recorded at selected intervals. The number of live fish showing pronounced symptoms of intoxication and distress, such as loss of equilibrium and other markedly abnormal behavioral anomalies should also be recorded. Dead fish must be removed as soon as they are observed.

Test Containers

While the size and shape of the test containers have not been standardized, the depth of the test water should be uniform in all vessels and at least 15 cm deep. The containers should be made of glass, except where more convenient materials have been shown to be nontoxic to fish and nonreactive with the test material (e.g., polyethylene adsorbs some organic substances, while glass adsorbs heavy metals).

Physical and Chemical Determinations

The determination of calcium and magnesium ions, total dissolved solids (TDS), alkalinity, temperature, dissolved oxygen, and pH of the test solutions and of measurable toxicant are normally made before introducing fish, and again, at the end of the test. This is also required for any other constituents of the water which markedly influence the toxicity of the material tested.

CALCULATION OF THE LC50

For each toxicity test, the LC50 and, if possible, the 95% confidence interval should be calculated on the basis of the volume percent of the effluent in the test solutions or the specific toxicant concentration. To allow a reasonable estimate of the LC50 and confidence interval to be made, the definitive test must meet the following criteria: (1) each concentration must be at least 50% of the next higher one, and (2) at least one concentration must have affected greater than 65% of the test organisms, while another (other than the control) must have affected less than 35% (Peltier, 1978; Parrish, 1984). A wide variety of manual methods are available to calculate a LC50 (Finney, 1971), including the log-concentration versus percent-survival and the Litchfield and Wilcoxon (1949) methods which are detailed below. Use of the Litchfield and Wilcoxon method requires partial mortality at dilutions above and below the LC50.

Frequently, in effluent toxicity tests, there is no partial mortality at any effluent concentration. In these cases, survival falls from 100% at one or more lower concentrations to 0% at the next higher concentration. When this occurs, it is not possible to calculate a confidence interval, and the log-concentration versus percent-survival method must be used.

Manual Methods for Estimating the LC50

Log-Concentration Versus Percent-Survival Method

(1) Plot the percent-effluent volumes and the corresponding percent-survival on semi-logarithmic paper (Figure 1).
(2) Locate the two highest points on the graph which are separated by the 50% survival line and connect them with a diagonal straight line. However, if one of the points is an aberrant value (as in Figure 1), the next lower or higher percent-effluent volume is used.
(3) Read on the scale for percent-effluent volume the value of the point where

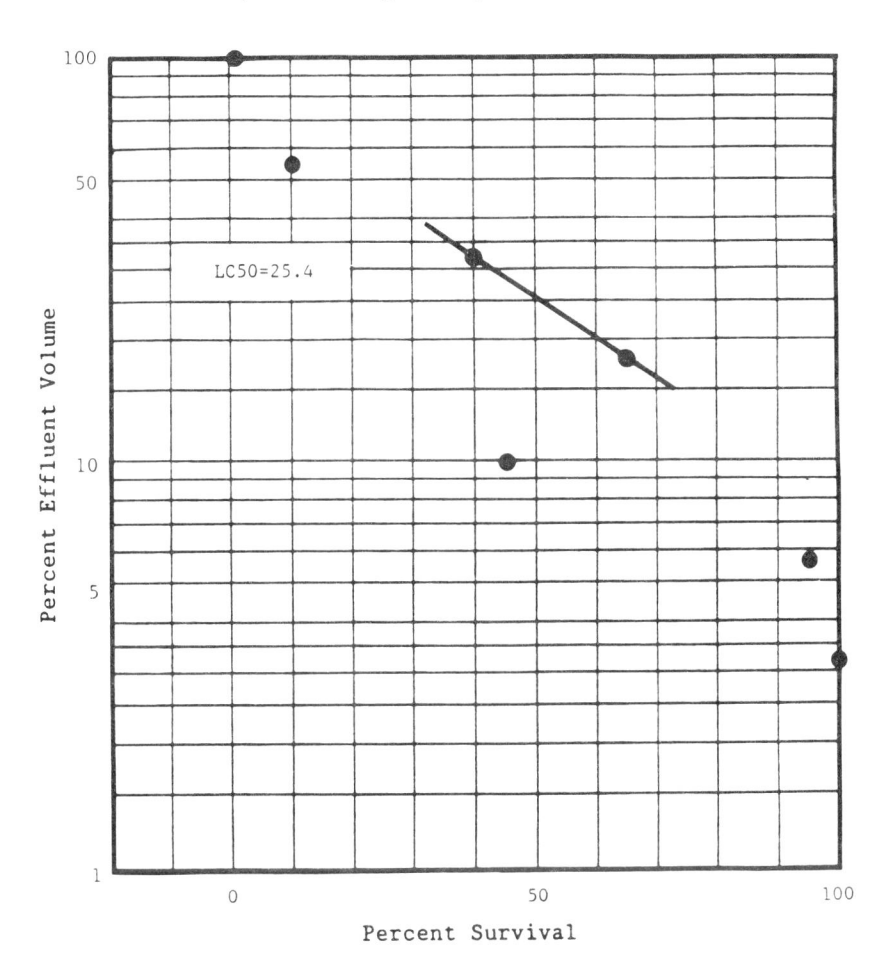

Aquatic Toxicology: Principles and Procedures 71

Figure 1 The plotted data and fitted line for the log-concentration versus percent-survival method of determining the LC50. From Plotkin and Ram (1983).

the diagonal line and the 50% survival line intersect. This value is the LC50, expressed as a percent-effluent volume, for the test.

Litchfield and Wilcoxon Abbreviated Method

(1) Tabulate the data showing the percent-effluent volumes used, the total number of organisms exposed to each percent-effluent volume, the number of affected organisms, and the observed percent-affected organisms. Do not list more than two consecutive 100% affects at the higher percent-effluent volumes or more than two consecutive 0% affects at the lower percent-effluent volumes.

(2) Plot the percent-affected organisms against the percent-effluent volume on two-cycle, logarithmic probability paper (Figure 2), except for 0% or 100% affect values. With a straight edge, fit a temporary line through the points, particularly those in the region of 40% to 60% affects.

(3) Using the line drawn through the points, read and list an expected percent affect for each percent-effluent volume tested. Percent-effluent volumes

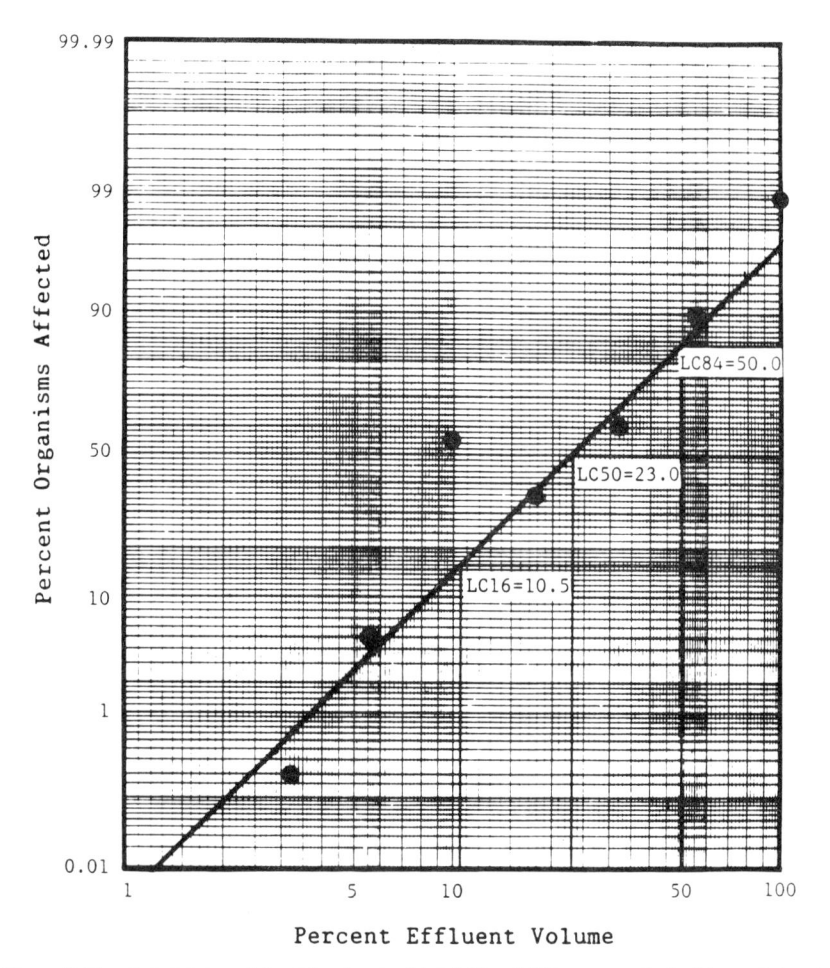

Figure 2 The plotted data and fitted line for the Litchfield and Wilcoxon method of determining the LC50. From Peltier (1978).

TABLE 1 Corrected values of 0 and 100 percent affects.

Expected Value	Corrected Value									
	0	1	2	3	4	5	6	7	8	9
0	—	0.3	0.7	1.0	1.3	1.6	2.0	2.3	2.6	2.9
10	3.2	3.5	3.8	4.1	4.4	4.7	4.9	5.2	5.5	5.7
20	6.0	6.2	6.5	6.7	7.0	7.2	7.4	7.6	7.8	8.1
30	8.3	8.4	8.6	8.8	9.0	9.2	9.3	9.4	9.6	9.8
40	9.9	10.0	10.1	10.2	10.3	10.3	10.4	10.4	10.4	10.5
50	—	89.5	89.6	89.6	89.6	89.7	89.7	89.8	89.9	90.0
60	90.1	90.2	90.4	90.5	90.7	90.8	91.0	91.2	91.4	91.6
70	91.7	91.9	92.2	92.4	92.6	92.8	93.0	93.3	93.5	93.8
80	94.0	94.3	94.5	94.8	95.1	95.3	95.6	95.9	96.2	96.5
90	96.8	97.1	97.4	97.7	98.0	98.4	98.7	99.0	99.3	99.7

with percent-affect values less than 0.01 and greater than 99.99 should be deleted from the list. Using the expected-percent affect, calculate from Table 1 a corrected value for each 0% or 100% affect obtained in the test. Since the expected values in the table are whole numbers, it will be necessary to obtain intermediate values by interpolation. Plot these values on the logarithmic probability paper (Figure 2) used in step two and inspect the fit of the line to the completely plotted data. If after plotting the corrected expected values for 0% and 100% affected, the fit is obviously unsatisfactory, redraw the line and obtain a new set of expected values.

(4) List the difference between each observed (or corrected) value and the corresponding expected value. Using each difference and the corresponding expected value, read and list the contributions to Chi-square (Chi^2) from Figure 3. A straight edge connecting a value on the expected-percent-affected scale with a value on the observed-minus-expected scale, will indicate at the point of intersection to Chi^2 scale, the contribution to Chi^2. Sum the contributions to Chi^2 and multiply the total by the average number of organisms per effluent volume, i.e., the number of organisms used in K concentrations divided by K, where K is the number of percent-affected organism values plotted. The product is the calculated Chi^2 of the line. The number of degrees of freedom (N) is equal to two less than the number of points plotted, i.e., $N = K - 2$. If the calculated Chi^2 is less than the Chi^2 given in Table 2 for N degrees of freedom, the data are non-heterogeneous and the line is a good fit. However, if the calculated Chi^2 is greater than the Chi^2 given in Table 2 for N degrees of freedom, the data are heterogeneous and the line is not a good fit. In the event a line cannot be fitted (the calculated Chi^2 is always

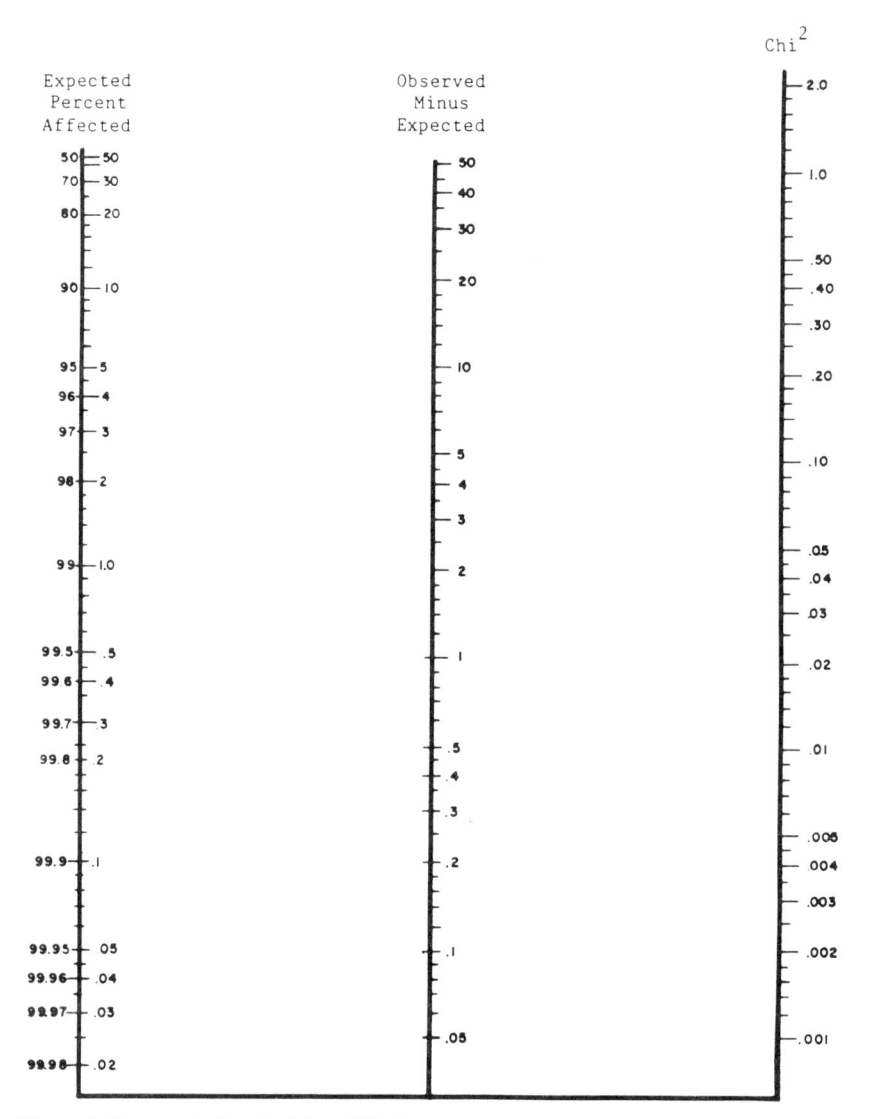

Figure 3 Nomograph for obtaining Chi² from expected-percent affected and observed-minus-expected values. From Peltier (1978).

TABLE 2 Values of Chi² ($P = 0.05$).

Degrees of Freedom (N)	Chi²
1	3.84
2	5.99
3	7.82
4	9.49
5	11.1
6	12.6
7	14.1
8	15.5
9	16.9
10	18.3

greater than the tabular Chi²), the data should not be used to calculate a LC50 or EC50. Litchfield and Wilcoxon (1949) provided an alternative method for calculating the 95% confidence limits under these circumstances. However, the toxicity test should be repeated.

(5) Read from the fitted line (Figure 2), the percent-effluent volumes for the corresponding 16, 50, and 84% affects (LC16, LC50, and LC84). Calculate the slope function, S, as:

$$S = \frac{LC84/LC50 + LC50/LC16}{2}$$

From the tabulation of the data determine N', the total number of test organisms used within the percent-affected organism interval of 16 to 84%. Calculate the exponent ($2.77/\sqrt{N'}$) for the slope function and the factor, f_{LC50}, which is used to establish the confidence limits for the LC50 (or EC50).

$$f_{LC50} = S^{(2.77/\sqrt{N'})}$$

The f_{LC50} can also be obtained directly from the nomograph in Figure 4 by laying a straight edge across the appropriate base and exponent values and reading the resultant f value. Calculate the 95% confidence limits of the LC50 as:

$$\text{upper limit} = LC50 \times f_{LC50}$$

$$\text{lower limit} = LC50/f_{LC50}$$

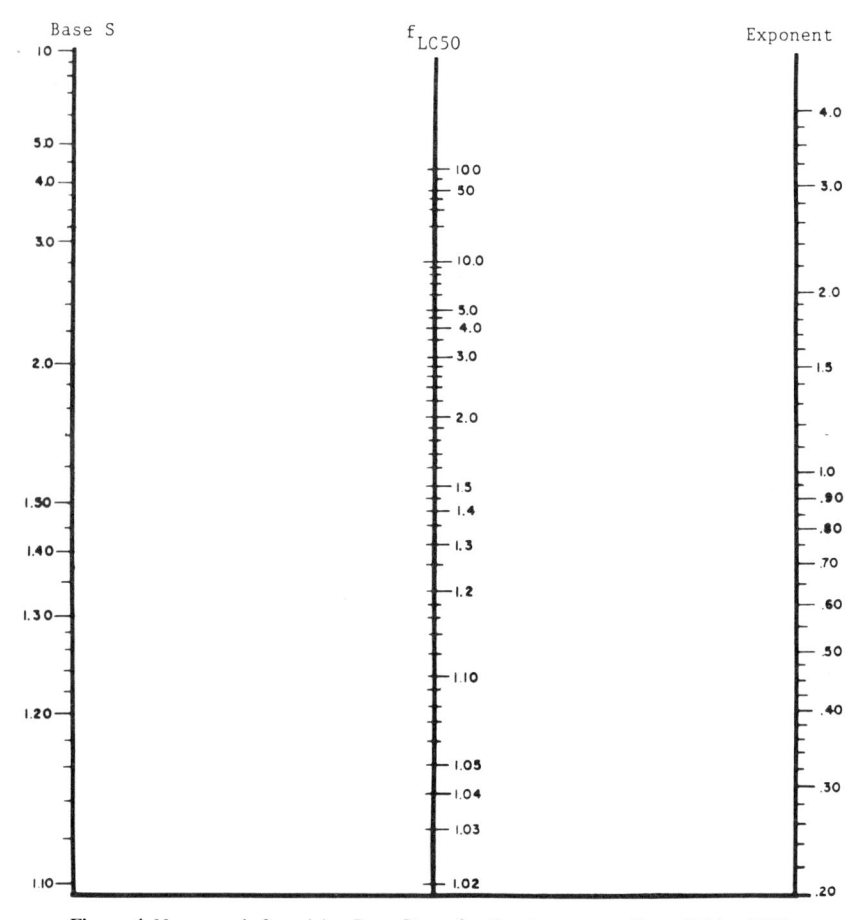

Figure 4 Nomograph for raising Base S to a fractional exponent. From Peltier (1978).

INCIPIENT LC50

The lethal threshold concentration or incipient LC50 (ILC50) is that level of toxicant below which 50% of the test specimens will not die from the stress factor upon prolonged exposure. Estimates may be obtained as follows: (1) Determine the median survival period (time required to cause 50% mortality at a given toxicant concentration) of the test organisms at several concentrations of toxicant. The median survival periods are then plotted on a log scale versus the log of the toxicant concentration. The asymptote to the time axis thus derived is the ILC50 approximately (Sprague, 1969). (2) The percent mortality is observed at several time periods throughout the test's duration. LC50 values for each of these exposure periods are interpolated using the previously described graphical procedure. The ILC50 is then evaluated from a log-log plot of the LC50 versus time. (3) A more accurate estimate of the ILC50, complete with confidence limits, may be made by selecting an exposure time beyond that where acute toxicity ceases and plotting the percent mortality (probability scale) versus toxicant concentration (log scale). The toxicant concentration corresponding to the 50% mortality value is identified as the ILC50.

TOXIC UNITS

The strength of a given toxicant may be expressed as a proportion of its lethal threshold concentration. Unity would represent the level corresponding to the ILC50. The numbers of this scale are designated as toxic units (Sprague and Ramsay, 1965). Because the strengths are expressed in the same units, a mixture of toxicants may be expressed as the sum of its toxic units.

This approach, while a convenient simplification, remains a generalization because it assumes that all effects are additive. Addition of concentrations low enough to have a minimal negative effect might result in a substantial overstatement of the resultant toxicity. Similarly, mixtures of toxicants affecting different physiological functions may produce a synergistically high value (Sprague, 1970). While most combinations do produce additive effects, this approach is limited by a severe lack of research in the prediction of the subtle, but equally profound sublethal effects.

APPLICATIONS OF LC VALUES

Much work has been done on extrapolation of acceptable field concentrations of test compounds from laboratory derived LC data. However, so many variables are involved that no satisfactory "fudge factor" has yet been evolved. Somewhat of a precedent was set at one location when pulping waste was tested at a 0.1 dilution of the LC50 value and results proved satisfactory to all concerned. Since then a basic rule of thumb used by many industries is to apply a 0.1 dilution factor to the LC50 value. Such generalizations have justifiably received criticism from many investigators, some claiming that a 0.01 dilution of the LC50 value is a more accurate estimate of the maximum allowable toxicant concentration (MATC). The relationship between the LC50 value and the MATC, if one exists, varies with the specific toxicant. One can generalize to generic types with only the greatest trepidation.

The assessment of a potential toxicant and the subsequent resolution of an application factor (AF) or its reciprocal, the chronicity value, involves the manipulation of two numbers generated by laboratory toxicity tests: (1) the 96-hour LC50 estimate of acute toxicity or the incipient LC50 (ILC50) estimate of the toxicity threshold and (2) the safe concentration or the MATC. The AF is derived by dividing the MATC by the 48 or 96-hour TLm or LC50 (Mount and Stephan, 1967) or the lethal threshold concentration or ILC50 (Eaton, 1970). Once determined, the AF permits estimation of MATC values for a given toxicant from just the 96-hour LC50 value.

While the generation of the latter requires no great competence, estimation of the MATC is a more difficult task. To define a threshold of inhibition for various life functions (growth, fecundity, etc.) over a sustained period demands a high level of expertise. For this reason, a current list of AF values would be extremely useful to laboratories evaluating sample toxicity.

The MATC can be determined directly by chronic toxicity studies. This determination involves placing fish fry in dilutions of the toxicant and observing growth inhibition and spawning interference. The MATC is determined from the geometric mean of the greatest concentration that causes the same response as the control (NOEC: no observed effect concentration) and the lowest concentration where a different response is observed (LOEC: lowest observable effect concentration). An alternative to determining an AF or MATC value based upon long-term chronic exposure is the estimation of "safe" levels using relatively short-term (30 to 60 day) exposures of eggs and larvae (Eaton, 1974).

Several factors should be considered in determining the MATC: (1) Long-term exposure of the test organisms to much lower toxicant concentrations than are lethal may still impair functions or performances such as swimming, appetite and growth, resistance to disease, reproductive capacity, and the general ability to compete with other species. For example, a certain concentration of toxicant may have no noticeable chronic effect on a particular species in the laboratory, but due to subtle impairment of some ability, the species might be completely eliminated from a waterway after a period of years. (2) Water quality parameters including alkalinity, hardness, pH, temperature, and conductivity can drastically change the LC50 (Sprague, 1984) and MATC under field conditions. (3) Low levels of a toxicant may have no effect on the species of fish tested but could have a toxic effect on other biota that the fish consume. Thus, certain species of fish might be selected against due to a change in food availability. (4) It is important to realize that the AF will change according to the toxicant and species being tested. Pickering and Gast (1972) demonstrated the AF for fathead minnows (*Pimephales promelas*) exposed to cadmium in hard water was 0.005 to 0.008 of the 96-hour LC50 value. The AF for the same species for copper was 0.03 to 0.08 of the 96-hour

LC50 (Mount, 1968). Eaton (1974) showed the AF for bluegill (*Lepomis machrochirus*) with cadmium as the toxicant to be 0.0015 to 0.0039 of the ILC50.

Some AF values have been compiled while many are still unknown. In the interest of expediting pollution control, a general AF must be applied when the AF or MATC values are unknown. AF values ranging from 0.01 to 0.1 of the 96-hour LC50 have been suggested (Petrocelli, 1984). Sprague (1971) proposed AF values of 0.05 to 0.1 of the 20 day LC50 (essentially the ILC50). Considering the findings that AF values vary greatly and may be well below 0.05 such as in the studies previously alluded to, a general AF value of 0.005 using the 96-hour LC50 may be a more reasonable value.

INTERPRETATION OF THE TOXICITY TEST

The applicability of experimental data has become an important topic in the field of fish toxicology. Numerous variables control the effects of toxicants upon fish as a whole. Such seemingly basic characteristics as temperature, dissolved oxygen, and pH are a few which can account for severalfold differences. Even more confusing is the combination of effects between a wide variety of toxicants and the physical and chemical characteristics of the total ecosystem. Yet, existing research has historically disregarded the dilution capacity and the assimilative nature of the receiving waters. Obviously, the precise evaluation of all effects is impossible. As in all real-life situations, decisions must be made from fragments of information and extrapolated to large-scale generalizations.

When extrapolating laboratory data to the field, one must remember that routine chronic laboratory experiments are predicated on a readily available food supply, though food is often limiting in the natural environment. Also, a decrease in the population or survival of fry may not result in a corresponding decrease in secondary productivity, whereas in the laboratory it most certainly will.

Consequently, biomonitoring and field verification techniques have been devised to circumvent the problem of creating artificially high or low pollution control standards in those instances where the combination of toxic effluents does not elicit an additive effect. Variable stream conditions attributable to production changes from predominantly daytime operations in industry and natural diurnal fluctuations have been largely ignored. Since, in natural aquatic systems, organisms are rarely exposed to stable concentrations, toxicity testing should, perhaps, include some procedures for a recovery period in toxicant-free water. Here again we are confronted with the possibility of creating overprotective control standards. One would expect certain combative body functions weakened by high concentrations of certain toxicants to be capable of replenishment during periods of low concentration. This may, in fact, necessitate varying standards for intermittent exposure based on the capacity for homeostatic reserve rejuvenation.

REFERENCES

APHA. *Standard Methods for the Examination of Water and Wastewater*, 16th ed. American Public Health Association, Washington, D.C., 1268 pp. (1985).

Brown, V. M. "Concepts and Outlook in Testing the Toxicity of Substances to Fish," in *Bioassay Techniques and Environmental Chemistry*, G. E. Glass, ed. Ann Arbor Science Publishers, Ann Arbor, pp. 73-95 (1973).

Doudoroff, P., B. G. Anderson, G. E. Burdick, P. S. Galtsoff, W. B. Hart, R. Patrick, E. R. Strong, E. W. Surber and W. M. Van Horn. "Bio-Assay Methods for the Evaluation of Acute Toxicity of Industrial Wastes to Fish," *Sewage Indust. Wastes*, 23:1380-1397 (1951).

Eaton, J. G. "Chronic Malathion Toxicity to the Bluegill (*Lepomis macrochirus* Rafinesque)," *Water Res.*, 4:673-684 (1970).

Eaton, J. G. "Chronic Cadmium Toxicity to the Bluegill (*Lepomis macrochirus* Rafinesque)," *Trans. Am. Fish. Soc.*, 103:729-735 (1974).

Finney, D. J. *Probit Analysis*, 3rd ed. Cambridge University Press, Cambridge, 333 pp. (1971).

Litchfield, J. T., Jr. and F. Wilcoxon. "A Simplified Method of Evaluating Dose-Effect Experiments," *J. Pharmac. Exp. Ther.*, 96:99-113 (1949).

Medeiros, C., R. Coler and N. Ram. *Principles and Procedures of Aquatic Toxicology*. National Training and Operational Technology Center, U.S. Environmental Protection Agency, Cincinnati, Ohio (1981).

Mount, D. I. "Chronic Toxicity of Copper to Fathead Minnows (*Pimephales promelas*, Rafinesque)," *Water Res.*, 2:215-223 (1968).

Mount, D. I. and W. A. Brungs. "A Simplified Dosing Apparatus for Fish Toxicology Studies," *Water Res.*, 1:21-29 (1967).

Mount, D. I. and C. E. Stephan. "A Method for Establishing Acceptable Toxicant Limits for Fish—Malathion and the Butoxyethanol Ester of 2,4-D," *Trans. Am. Fish. Soc.*, 96:185-193 (1967).

Parrish, P. R. "Acute Toxicity Tests," in *Fundamentals of Aquatic Toxicology*, G. M. Rand and S. R. Petrocelli, eds. Hemisphere Publishing Co., Washington, pp. 31-57 (1984).

Peltier, W. *Methods for Measuring the Acute Toxicity of Effluents to Aquatic Organisms*. Environmental Monitoring and Support Laboratory, U.S. Environmental Protection Agency, Cincinnati, Ohio (1978).

Peltier, W. H. and C. I. Weber. *Methods for Measuring the Acute Toxicity of Effluents to Freshwater and Marine Organisms*, 3rd ed. Environmental Monitoring and Support Laboratory, U.S. Environmental Protection Agency, Cincinnati, Ohio (1985).

Peterson, R. H. and J. M. Anderson. "Influence of Temperature Change on Spontaneous

Locomotor Activity and Oxygen Consumption of Atlantic Salmon, *Salmo salar*, Acclimated to Two Temperatures," *J. Fish. Res. Bd. Can.*, 26:93–109 (1969).

Petrocelli, S. R. "Chronic Toxicity Tests," in *Fundamentals of Aquatic Toxicology*, G. M. Rand and S. R. Petrocelli, eds. Hemisphere Publishing Co., Washington, pp. 96–109 (1984).

Pickering, Q. H. and M. H. Gast. "Acute and Chronic Toxicity of Cadmium to the Fathead Minnow (*Pimephales promelas*)," *J. Fish. Res. Bd. Can.*, 29:1099–1106 (1972).

Plotkin, S. and N. M. Ram. *Acute Toxicity Tests: General Description and Materials and Methods Manual I. Fish.* Environmental Engineering Program, Department of Civil Engineering, University of Massachusetts, Amherst (1983).

Sprague, J. B. "Measurement of Pollutant Toxicity to Fish I. Bioassay Methods for Acute Toxicity," *Water Res.*, 3:793–821 (1969).

Sprague, J. B. "Measurement of Pollutant Toxicity to Fish II. Utilizing and Applying Bioassay Results," *Water Res.*, 4:3–32 (1970).

Sprague, J. B. "Measurement of Pollutant Toxicity to Fish III. Sublethal Effects and 'Safe' Concentrations," *Water Res.*, 5:245–266 (1971).

Sprague, J. B. "Factors that Modify Toxicity," in *Fundamentals of Aquatic Toxicology*, G. M. Rand and S. R. Petrocelli, eds. Hemisphere Publishing Co., Washington, pp. 124–163 (1984).

Sprague, J. B. and B. A. Ramsay. "Lethal Levels of Mixed Copper-Zinc Solutions for Juvenile Salmon," *J. Fish. Res. Bd. Can.*, 22:425–432 (1965).

Stephan, C. E. "Methods for Calculating an LC50," in *Aquatic Toxicology and Hazard Evaluation*, F. L. Mayer and J. L. Hamelink, eds. ASTM SP 634, American Society for Testing and Materials, Philadelphia, pp. 65–84 (1977).

EXERCISE 10/TOXICITY TESTING WITH FISH[2]

PURPOSE

The purpose of this exercise is to demonstrate some of the techniques used to determine the acute toxicity (96-hour LC50) of toxicants to fish. The incipient LC50 and maximum allowable toxicant concentration of the test site water will also be estimated.

INTRODUCTION

Three steps are required to evaluate the toxicity of a sample: (1) the determination of the range of toxicant which results in an observable response, (2) the determination of the 96-hour LC50 and the time-independent or incipient

---

[2]Taken, in part, from Medeiros et al., 1981.

LC50 (ILC50), and (3) the calculation of the maximum allowable toxicant concentration (MATC).

The first step is a qualitative screening procedure used to assign a range of toxicity. One such procedure is the ranging oxygen test. Several fish are placed into BOD bottles containing aerated control (dilution) water or control water plus some level of toxicant. The bottles are then sealed and left until 100% fish mortality is observed. The dissolved oxygen (DO) concentration in each bottle is determined immediately after 100% mortality is obtained. The DO values in the bottles in which fish mortality is attributable to oxygen depletion alone will be on the order of tenths of a mg O_2/l. DO values in the bottles in which fish mortality is attributable to the toxicant will be higher since mortality will occur prior to oxygen depletion. Thus, DO levels in the bottles containing higher toxicant concentrations will be greater than in those containing lower toxicant concentrations. If the residual DO concentration is plotted against the log of the toxicant concentration, an inflection point is observed where the DO increases with increasing toxicant level. Toxicant levels just below and above this point are then tested over a finer incremental dosage range for the 96-hour LC50 determination. The incremental range is distributed over a logarithmic interval owing to the exponential response of test organisms to toxicant dosage (Gaddum, 1953).

Alternatively, a 24-hour screening jar test can be conducted over a wide range of toxicant levels. This test is usually performed using ten-fold dilutions of the toxicant and observing the mortality of four or five test organisms at these toxicant levels after a 24-hour period. The finer incremental toxicant dosage range to be used for the 96-hour LC50 toxicity test is then determined.

The second stage of evaluation is the determination of the 96-hour LC50 and, if possible, the ILC50. This is a static (or static with daily renewal) test performed with two replicates per concentration and ten fish per aquarium. The range of toxicant concentrations required to observe the 96-hour LC50 value is determined using one of the qualitative screening procedures described above.

The final evaluation of the toxicity of the sample is the MATC. The MATC is bounded by the highest concentration that elicits the same response as the control and the lowest test concentration having a significantly different effect in a life cycle or a partial life cycle test (Petrocelli, 1984). The MATC can be approximated by multiplying the ILC50, or if it is not available, the 96-hour LC50, by a chemical specific application factor (AF). Hypothetical AF values ranging from 0.05 to 0.1 (Sprague, 1971) and from 0.01 to 0.1 (Petrocelli, 1984) have been used for this purpose.

PROCEDURE

Acclimation and Handling

Acclimate the fish to the experimental conditions (light, temperature, etc.) in control site or dilution water for a minimum of 14 days (APHA, 1985). During this period maintain them on a daily diet of commercially available foods (Peltier and Weber, 1985). Feeding should be interrupted two days prior to the start of and during static tests of limited duration (96 hours or less). Remove abnormal and dead individuals from the holding tank and if mortality exceeds 10%, discard the population (APHA, 1985). Further, mortality in excess of 5% during the 24 to 48 hours immediately prior to a test is unacceptable (Peltier and Weber, 1985).

Avoid the unnecessary handling of fish. To minimize stress, transfers should be made as gently and as rapidly as possible using a dip-net for the larger fish and a wide-bore (4 to 8 mm) pipet with a rubber bulb for larvae (Peltier and Weber, 1985). Fish that are dropped or injured during the transfer process are not to be used in toxicity tests.

Screening Tests

Ranging Oxygen Test

Prepare solutions consisting of 0.001, 0.01, 0.1, 1, 10, and 100% test site water. A control must be included for comparative purposes. Transfer the test solutions to duplicate 300 ml BOD bottles for each concentration, place five fish (2 to 2.5 cm in length) into each bottle, and carefully stopper to avoid trapping air bubbles. Record the starting time in Table 10-1 and leaving the bottles undisturbed, observe the fish at 20-minute intervals. Immediately following the death (cessation of gill movement) of the last fish in each bottle, note the time and determine its DO concentration using the Alsterberg azide modification of the Winkler method (APHA, 1985). Use care as the DO levels in the control bottles will be extremely low. Record the results in Table 10-1.

Plot, on semi-logarithmic paper, the toxicant concentration (on log scale) versus the residual DO level. Determine the inflection point and choose a narrower range around this point for use in the 96-hour test. Alternatively, the data can be plotted on log-log paper (Vigers and Maynard, 1977). With this option the data are segregated into two groups and a best fit line is drawn

TABLE 10-1 Ranging oxygen test results.

Test Concentration		Start Time	Finish Time	Duration (hours)	DO Level (mg O_2/l)	Comments
Control	1					
	2					
0.001%	1					
	2					
0.01%	1					
	2					
0.1%	1					
	2					
1%	1					
	2					
10%	1					
	2					
100%	1					
	2					

through each set. The first line, consisting of the control and lower toxicant concentrations, has a slope of zero. The second line has a positive slope and is referred to as the response line. The intersection of the two lines is used, like the inflection point above, to determine the range of toxicant concentrations for the definitive test.

Screening Jar Test

Prepare three liters of the solutions (as above) and pour them into individual one-gallon jars. A control containing no toxicant must also be used. Carefully place five fish of the same species, age, size, and sex into each container. Observe the percent mortality after 24 hours. Determine the highest toxicant concentration resulting in 0% mortality and the lowest resulting in 100% mortality. A narrower toxicant range lying between these points will be used in the 96-hour test.

Definitive Test

The procedure for the 96-hour LC50 determination using a static system with daily renewal is essentially the same as that described for the screening jar test. The required concentrations (a minimum of five at equal logarithmic intervals) are prepared and poured into two test containers per concentration. Also prepare duplicate control jars. The fish are then captured from a common tank and distributed sequentially until ten fish are placed into each container (Peltier and Weber, 1985). To avoid oxygen depletion, loading should not exceed 0.5 to 0.8 g of fish per liter of test solution (Parrish, 1984). Renew the test solutions daily.

Observe the fish regularly, after 0.5, 1, 3, 6, 12, 24, 48, 72, and 96 hours. Record in Table 10-2, the percent mortality, as well as any additional observations such as discoloration, mucus coagulation, agitation, twitching, turnover, or other behavioral anomalies. Remove dead fish from the test vessels. If control death exceeds 10%, repeat the test using a different batch of fish.

Determine the 96-hour LC50 with confidence limits (if possible) using one of the previously described procedures and record the data in Table 10-2. For each concentration plot, on three-cycle, logarithmic probability paper, the observation time versus percent mortality. Determine the median survival time for each concentration by interpolating to 50% mortality. Record these values in Table 10-2. Plot, on log-log paper, the concentration versus the median survival time. The asymptote to the time axis approximately represents the ILC50 (Sprague, 1969). Record in Table 10-2. Also, determine the MATC using the AF ranges provided above and record them in Table 10-2, indicating which range was used.

QUESTIONS

(1) Discuss the advantages and disadvantages of the ranging oxygen test.
(2) In the ranging oxygen test, why doesn't it matter if the size, sex, and number of fish in each replicate and concentration are not controlled?
(3) Suppose you are an employee in a state water pollution control laboratory. How would you derive a MATC for lead contamination without the resources to implement a chronic toxicity test?
(4) Suppose your test water is acutely toxic. How would you identify the causative factor(s)?
(5) What would be the advantages of running two or more toxicity tests on

TABLE 10-2 **Acute toxicity of test site water to fish.**

Test Concentration		Percent Mortality at Time (hours)								Median Survival Time	Comments
		0.5	1	3	6	12	24	48	72	96		
Control	1											
	2											
	1											
	2											
	1											
	2											
	1											
	2											
	1											
	2											
	1											
	2											
	1											
	2											
	1											
	2											
	1											
	2											
	1											
	2											
	1											
	2											

96-Hour LC50:

ILC50:

MATC:

the same sample with representative organisms from different kingdoms or phyla?

REFERENCES

APHA. *Standard Methods for the Examination of Water and Wastewater*, 16th ed. American Public Health Association, Washington, D.C., 1268 pp. (1985).

Gaddum, J. H. "Bioassays and Mathematics," *Pharmac. Rev.*, 5:87–134 (1953).

Medeiros, C., R. Coler, and N. Ram. *Principles and Procedures of Aquatic Toxicology*. National Training and Operational Technology Center, U.S. Environmental Protection Agency, Cincinnati, Ohio (1981).

Parrish, P. R. "Acute Toxicity Tests," in *Fundamentals of Aquatic Toxicology*, G. M. Rand and S. R. Petrocelli, eds. Hemisphere Publishing Co., Washington, pp. 31–57 (1984).

Peltier, W. H. and C. I. Weber. *Methods for Measuring the Acute Toxicity of Effluents to Freshwater and Marine Organisms*, 3rd ed. Environmental Monitoring and Support Laboratory, U.S. Environmental Protection Agency, Cincinnati, Ohio (1985).

Petrocelli, S. R. "Chronic Toxicity Tests," in *Fundamentals of Aquatic Toxicology*, G. M. Rand and S. R. Petrocelli, eds. Hemisphere Publishing Co., Washington, pp. 96–109 (1984).

Sprague, J. B. "Measurement of Pollutant Toxicity to Fish I. Bioassay Methods for Acute Toxicity," *Water Res.*, 3:793–821 (1969).

Sprague, J. B. "Measurement of Pollutant Toxicity to Fish III. Sublethal Effects and 'Safe' Concentrations," *Water Res.*, 5:245–266 (1971).

Vigers, G. A. and A. W. Maynard. "The Residual Oxygen Bioassay: A Rapid Procedure to Predict Effluent Toxicity to Rainbow Trout," *Water Res.*, 11:343–346 (1977).

EXERCISE 11/ALGAL TOXICITY TESTING IN A FLOW-THROUGH GLASS COIL ASSEMBLY

PURPOSE

This session will be devoted to evaluating the toxicity of test site water using flow-through glass coil assemblies colonized by control site periphyton. The EC50 will be derived based on the test water's inhibitory effect on photosynthesis as indicated by oxygen generated within the glass coils. The student will recognize this protocol as a modification of the field study of periphyton productivity in Exercise 9.

INTRODUCTION

The use of algae as test organisms in toxicity testing procedures, to predict and evaluate nutrient and toxicant effects on eutrophication rates and trophic status, is well documented. Various methods and procedures structured to fit specific situations have been reported, but these have offered no basis for comparison between water bodies. Consequently, in 1968 a group of scientists carefully scrutinized the tentative procedure set up by the National Eutrophication Research Program in response to the need for a standardized algal growth test. In 1969 the Provisional Algal Assay Procedure (PAAP) was published describing three alternative procedures (Joint Industry/Government Task Force on Eutrophication, 1969). After two years of comparison, the Algal Assay Procedure: Bottle Test (AAP) was adopted as the preferred method (USEPA, 1971).

Current algal assays, such as the AAP, utilize one or several species of test organisms which provide a range of responses to the nutritional and physical characteristics of the test water (Payne, 1975; Parker, 1977). In these tests, the growth rates of the algae are monitored in control and test flasks which are maintained on a shaker with an artificial light source for 14 days. Periodic analysis of the chlorophyll *a* content or biomass over the test period indicates the growth rate of the algae in various concentrations of the test solution. Allowable concentrations of the toxicant or limiting nutrients are derived from comparisons of test growth rates with that of the control.

The application of this procedure to streams and fast-flowing rivers, however, where the algal community is dominated by periphyton, is somewhat tendentious. The AAP method is predicated on a confined, nonflowing system, which, while satisfactory for planktonic forms, is inappropriate for the requirements of periphyton. Paradoxically, though our rivers have received the greatest burden of our burgeoning population and industrialization, algal assays using periphyton have not been developed to the extent of the static AAP method (Weber, 1973).

Of the handful of investigators studying the periphytic response in the laboratory, the contribution of Odum and Hoskin (1957) is noteworthy. They measured the metabolic rate of a periphyton community by monitoring the change in oxygen concentration through time in a flowing water microcosm constructed from a glass condenser tube. An open trough methodology in which the biomass, productivity, and community structure of the resident community were measured was developed by McIntire and Phinney (1965). More recently, Trotter and Hendricks (1979) used chlorophyll and biomass measurements to infer productivity rates of attached algal mats exposed to various test waters in continuous-flow systems constructed from battery jars. A gravity feed flow-through system was used by Tease and Coler (1984) to measure, through changes in oxygen levels, the productivity of periphyton colonizing glass coils.

This exercise incorporates a flow-through system in which periphyton colonize the inner walls of vertically hung glass coils fed by reservoirs containing various dilutions of test water. Unlike the AAP, the flow-through system infers the rate of primary productivity from the oxygen generated by the periphytic algae within the coil. This model then affords a more rational extrapolation to lotic waters than the use of the planktonic model of the AAP.

PROCEDURE

Setup and Colonization

Six coiled glass tubing units (tubing 4 mm inside diameter and 366 cm in length; coils 10 cm in diameter and 50 cm in length) are suspended vertically from ring stands flanking a fluorescent light source. A 20 liter reservoir is positioned above each coil to provide a gravity feed flow-through system. Plastic tubing is used to connect the reservoir to the coil and the coil to the drain. Adjustable pinch clamps on the tubing above the coil are used to regulate the flow rate.

The carboys are filled with water from the control site which is permitted to flow through the coils for 10 to 14 days allowing for colonization of the tubes by periphyton. Throughout the colonization and experimental periods, a continuous light source of 400 ft-candles and a flow rate of 5 to 10 ml/minute are maintained. The same flow rate should be used for all of the systems and should be maintained for the duration of the experiment. Also, every effort must be made to keep the temperature constant, since the metabolism of algae roughly doubles for each 10°C rise (Eppley, 1972).

Baseline and Experimental Productivity

Once the coils have been colonized evenly throughout and a minimum dissolved oxygen (DO) differential between the carboy and outflow of 0.5 mg O_2/l recorded, the assay can proceed. It is desirable for all of the coils to have comparable productivity values, that is, equivalent biomass.

While the control site water is flowing through all of the coils, determine the baseline productivity using the sampling procedure described below. Record the data in the top portion of Table 11-1. Replace the water in the carboys with the appropriate concentration of test water [0 (control), 6.25, 12.5, 25, 50, and 100%] and determine the experimental productivity rates at 24-hour intervals for a minimum of four days. Record these values in the bottom section of Table 11-1.

TABLE 11-1 **Baseline and experimental productivity.**

Date Hour Temp	Toxicant Level	Carboy DO			Outflow DO			DO Differential (mg O$_2$/l)	Flow (l/min)	Net Productivity (mg O$_2$/min)
		1	2	$\bar{x}$	1	2	$\bar{x}$			
	Baseline									
	Baseline									
	Baseline									
	Baseline									
	Baseline									
	Baseline									
	Control									
	6.25%									
	12.5%									
	25%									
	50%									
	100%									
	Control									
	6.25%									
	12.5%									
	25%									
	50%									
	100%									
	Control									
	6.25%									
	12.5%									
	25%									
	50%									
	100%									
	Control									
	6.25%									
	12.5%									
	25%									
	50%									
	100%									

The net productivity for the coil (mg O_2/minute) is determined by multiplying the DO differential (mg O_2/l) by the flow rate (l/minute). Plot the toxicant concentration versus the mean productivity rates. The EC50 can be estimated by interpolating to the concentration reducing algal productivity by 50%.

Sampling Procedure

Using a 10 ml volumetric flask and a stopwatch, determine the flow rate (l/minute) and record the results in Table 11-1. Collect duplicate samples for each coil in 60 ml DO bottles by propping the bottle up under the outflow until the tubing is near the bottom of the bottle. Allow it to overflow for a period of time at least equal to the time required to fill the bottle. Carefully remove the bottle from underneath the outflow and stopper it with care to avoid trapping air bubbles. Similarly, collect duplicate samples from the carboy. Determine the DO concentrations using the Alsterberg azide modification of the Winkler method (APHA, 1985) using 0.5 ml of each reagent and one-half strength titrant. When 50 ml is titrated, the DO concentrations (mg O_2/l) are equal to the ml of titrant used multiplied by 2.

QUESTIONS

(1) How do temperature and current affect production?
(2) For what reasons should a single algal species be used per coil rather than a mixture? Why should a variety of species be exposed to the same toxicant before estimating an EC50?
(3) The Q_{10} law for temperature states that for each 10°C increase the metabolic rate doubles. How would you determine a Q_{10} for current?
(4) Describe an experimental design that would incorporate an assessment of the treatment impact on the composition and diversity of a community.
(5) How would you derive a productivity/respiration (P/R) ratio?

REFERENCES

APHA. *Standard Methods for the Examination of Water and Wastewater*, 16th ed. American Public Health Association, Washington, D.C., 1268 pp. (1985).

Eppley, R. W. "Temperature and Phytoplankton Growth in the Sea," *Fish. Bull.*, 70:1063–1085 (1972).

Joint Industry/Government Task Force on Eutrophication. *Provisional Algal Assay Procedure*. P.O. Box 3011, Grand Central Station, New York, NY 10017, 62 pp. (1969).

McIntire, C. D. and H. K. Phinney. "Laboratory Studies of Periphyton Production and Community Metabolism in Lotic Environments," *Ecol. Monogr.*, 35:237–258 (1965).

Odum, H. T. and C. M. Hoskin. "Metabolism of a Laboratory Stream Microcosm," *Publ. Inst. Mar. Sci. Univ. Texas*, 4:115-133 (1957).

Parker, M. "The Use of Algal Bioassays to Predict the Short- and Long-Term Changes in Algal Standing Crop Which Result from Altered Phosphorus and Nitrogen Loadings," *Water Res.*, 11:719-725 (1977).

Payne, A. G. "Responses of the Three Test Algae of the Algal Assay Procedure: Bottle Test," *Water Res.*, 9:437-445 (1975).

Tease, B. and R. A. Coler, "The Effect of Mineral Acids and Aluminum from Coal Leachate on Substrate Periphyton Composition and Productivity," *J. Freshw. Ecol.*, 2:459-467 (1984).

Trotter, D. M. and A. C. Hendricks. "Attached, Filamentous Algal Communities," in *Methods and Measurements of Periphyton Communities: A Review*, R. L. Weitzel, ed. ASTM STP 690. American Society for Testing and Materials, Philadelphia, pp. 58-69 (1979).

USEPA. *Algal Assay Procedure Bottle Test*. National Eutrophication Research Program, Pacific Northwest Water Laboratory, Corvallis, Oregon (1971).

Weber, C. I. *Biological Field and Laboratory Methods for Measuring the Quality of Surface Waters and Effluents*. Environmental Monitoring Series, U.S. Environmental Protection Agency, Cincinnati, Ohio (1973).

EXERCISE 12/TOXICITY TESTING WITH *DAPHNIA*

PURPOSE

The EPA recommends the use of *Daphnia* in toxicity testing on effluents rather than an indigenous species. The use of *Daphnia* provides a uniform approach that allows comparison and correlation of results. Furthermore, the size of the organism permits the use of large numbers in the test vessels which gives a more accurate statistical result. Their ease in culturing, short lifecycle, and ready availability allow for an almost inexhaustible supply. Such attributes provide an economy of time and space and thus *Daphnia* are extremely cost effective.

INTRODUCTION

Daphnia, as freshwater organisms, are ubiquitous in all but lotic and polluted environments. Most species are eurythermal and the range of unpolluted habitats for these sensitive organisms is governed primarily by pH, organic carbon content, hardness, and dissolved oxygen (Pennak, 1978). In the trophic pyramid of an aquatic system, *Daphnia* serve as an important link between primary producers and saprophytic bacteria on one side and higher consumers such as rotifers and fish on the other (Mount and Norberg, 1984).

The population is seasonal and, depending on the species and the habitat, is either mono- or dicyclic. The first sign of increasing population occurs seasonally when waters attain temperatures of 6 to 12°C. A peak is reached in mid to late April, after which the density dwindles owing to high temperatures, changing food conditions, and increases in invertebrate and vertebrate predation. If the population is dicyclic, a rebound occurs in autumn. The population density then declines to a low in winter with little or no reproduction evident (Pennak, 1978).

For the most part of the year, the population is dominated by parthenogenetic females. Parthenogenesis, as a form of asexual reproduction, provides a low level of genetic variability. The induction of males is the result of deteriorating environmental conditions such as high population densities, an increasing accumulation of excretory products, or a decreasing food supply (Weber and Peltier, 1981). If adverse conditions continue, sexual eggs will be produced. Instead of forming parthenogenetic clutches of ten to twenty individuals, the sexual female will produce only a few resting eggs, enclosed in an ephippium (Pennak, 1978). Upon molting, the ephippium is released to float to the surface and be dispersed by the wind or sink to the sediments where the eggs hatch when environmental conditions become suitable.

The development of *Daphnia* has been divided into four stages: egg, juvenile, adolescent, and adult (Pennak, 1978). In each instar, growth occurs immediately following the molt before the new carapace hardens (Weber and Peltier, 1981). In *D. magna*, the average juvenile stage consists of three to five instars while the adolescent period consists of a single instar. In this phase the first clutch of eggs reaches full development. The daphnid enters the first adult instar when the eggs are released into the brood chamber. The adult phase consists of six to twenty-two instars, each characterized by the release of hatched, first-instar juveniles from a brood chamber, a molt, size enlargement, and a release of new clutch into the brood chamber (Pennak, 1978, Weber and Peltier, 1981). The number of young released per brood is influenced positively by the nutrient content of the water. The frequency of reproduction may increase with increasing temperature, light, and food concentrations while life span increases with decreasing population density and declining temperature (Weber and Peltier, 1981).

PROCEDURE

Culturing

Daphnids may be cultured in either unpolluted surface or groundwater, dechlorinated tap water, or in reconstituted water. Moderately hard (80 to 90 mg $CaCO_3$/l for *D. pulex*) or hard (160 to 180 mg $CaCO_3$/l for *D. magna*) reconstituted water can be prepared by dissolving 96.0 or 192.0 mg NaHCO$_3$/l, 60.0 or 120.0 mg CaSO$_4$·2H$_2$O/l, 60.0 or 120.0 mg MgSO$_4$/l, and 4.0 or 8.0 mg KCl/l, respectively in deionized or distilled water (Weber and Peltier, 1981). These solutions should be aerated vigorously for several hours prior to use. It is best if the daphnids are cultured in control site water under the conditions (light, temperature, etc.) that will be used during experimentation.

Although two to four-liter wide-mouth glass jars are frequently used for cultures, five-gallon glass aquaria may be preferred because of the greater surface area to volume ratio which obviates the need for aeration. To avoid the loss of an entire daphnid colony, the investigator should maintain several (minimum of five) culture vessels and thin them regularly on a staggered basis. The containers should be covered with glass or plastic to minimize evaporation and exclude foreign debris.

The culturing of *Daphnia* can be successful over a wide range of temperatures. The optimum temperature is roughly 20°C, and if laboratory temperatures remain in the range of 18 to 26°C, normal growth and reproduction can occur without the need for costly temperature control equipment (Weber and Peltier, 1981). Also, at least sixteen hours of light (50 to 100 ft-candles) per day should be provided.

Appropriate food preparation and feeding are very important in maintaining daphnid cultures. Weber and Peltier (1981) recommend a suspension of trout chow, alfalfa, and dried yeast. This is prepared by blending 6.3 g of trout chow pellets, 2.6 g of dried yeast, and 0.5 g of dried alfalfa with 500 ml of distilled water at high speed for five minutes. The resulting mixture is placed into a refrigerator and allowed to settle for one hour. After it has settled, decant 300 ml of the supernate and pour 30 to 50 ml aliquots into 100 ml polyethylene, screw-top bottles and freeze. The portions should be thawed as needed. The food solution should then be kept refrigerated and if not used within the week, it should be discarded.

Approximately 1.5 ml of food solution per liter of culture medium should be added three times per week. Cultures will crash if they are fed at less frequent intervals. Small amounts of excess food do not pose a problem provided the solution is continuously aerated and replaced every week. Infrequent replacement of the culture media will result in the accumulation of waste products which may lead to a population crash or to the production of males and ephippia (Weber and Peltier, 1981). Cultures containing ephippia should not be used for toxicity testing.

Concurrently, the daphnid culture should be thinned to prevent overcrowding. Medium replacement and thinning are accomplished by gently transferring 10 to 15 adult *Daphnia* per liter of culture medium to a beaker containing roughly 100 ml of fresh solution. The handling and transferring of *Daphnia* is achieved through the use of a disposable pipet whose tip has been cut and fire polished such that the diameter of the opening is roughly 5 mm. A pipet bulb provides sufficient suction to entrain the organisms. During the transfer, the tip of the pipet should be kept beneath the surface of the water to avoid air entrapment under the daphnid carapaces. Discard the remaining *Daphnia* and the old culture solution and clean the vessel with detergent. The vessel should be rinsed thoroughly with the tap water followed by five rinses with distilled water. Fill the vessel with freshly prepared medium, gently pour the daphnids from the beaker to the vessel, and replace the plastic or glass cover.

Toxicity Testing

Test Organisms

First instar *D. magna* (<24 hours old) should be used for conducting toxicity tests. The required number can be obtained by transferring 10 adult females bearing embryos in their brood chamber to each of five 500 ml beakers containing 300 ml of culture medium and 0.5 ml of food solution 24 hours prior to the start of the test (Weber and Peltier, 1981). The young found the following day in the beakers are then used for the toxicity test.

Standard Toxicant

To assure that a healthy and normal *Daphnia* culture has been established in the laboratory, a toxicity test using a known standard of toxicant, such as sodium dodecyl sulfate (SDS), should be conducted. The 48-hour LC50 value for this toxicant for *D. magna* should fall between 5 and 10 mg SDS/l for water with a hardness of 160 to 180 mg $CaCO_3$/l (see Peltier and Weber, 1985). If the 48-hour LC50 does not fall in the recommended range, the test procedure is suspect and the tests, with both SDS and the test site water, should be repeated using a different batch of organisms.

In this test, a minimum of three reference toxicant concentrations (5, 10, and 15 mg SDS/l), in addition to a control, should be employed. Follow the definitive test procedure described below utilizing two replicate dishes per concentration, each containing five *Daphnia*. Determine the mortality after 48 hours and estimate the 48-hour LC50 for SDS.

Acute Toxicity Tests

Two steps will be used to evaluate the acute toxicity of the test site water to *Daphnia*: (1) The screening test is used to determine the range of toxicant dosage which results in an observed response. (2) The definitive test determines the acute toxicity using toxicant concentrations applied over a narrower range. It is the definitive test which is used to determine the 48-hour LC50.

A 24-hour screening test is first initiated on the test site water to determine the overall toxicity range. In this test, the concentrations are varied by a factor of ten, ranging from 0.001 to 100% test site water. Use one test vessel containing five *Daphnia* per concentration (including a control) and follow the test procedure described below. Determine the mortality for each concentration after 24 hours.

The definitive test concentrations are determined by dividing logarithmically the interval between the highest concentration that produced 0% mortality and the lowest concentration that caused 100% mortality in the screening test. For example, if six concentrations are required (in addition to the control) and the previously mentioned values are 10 and 100%, respectively, the concentrations to be used in the test are 10, 16, 25, 40, 63, and 100% test water.

The vessels used for testing will be petri dishes. Their large surface area, in conjunction with a light source below, will help in censusing the subjects. To each of four replicate vessels per concentration, add 50 ml of the appropriate concentration of test or control site water and five first instar *Daphnia*. No food should be added to the dishes, and to reduce carryover of materials (food, wastes, etc.) from the culture vessel, the daphnids should be carefully transferred through three changes of control site water (five minutes per wash) prior to being placed in their test vessels (APHA, 1985). The daphnids are transferred with a pipet, as described earlier, being careful to avoid stressing the organism which could increase its sensitivity to the toxicant. If during these initial transfers an individual should be confined to the surface film by an air bubble trapped beneath its carapace, it should be replaced.

The *Daphnia* should be observed regularly, after 1, 2, 4, 8, 16, 24, and 48 hours. During each observation, the number of nonmotile individuals in each test vessel should be recorded in Table 12-1. An individual is considered to be nonmotile, but not necessarily dead, if it shows no independent movement (antennae, thoracic appendages, or postabdomen) after the vessel has been rotated (Buikema et al., 1980; APHA, 1985). Using these data, determine the 48-hour LC50 and the ILC50.

TABLE 12-1 Acute toxicity of test site water to *Daphnia*.

Test Concentration		Nonmotile *Daphnia* at Time (hours)							Comments
		1	2	4	8	16	24	48	
Control	1								
	2								
	3								
	4								
	1								
	2								
	3								
	4								
	1								
	2								
	3								
	4								
	1								
	2								
	3								
	4								
	1								
	2								
	3								
	4								
	1								
	2								
	3								
	4								
	1								
	2								
	3								
	4								
	1								
	2								
	3								
	4								

Chronic Toxicity Test and Reproductive Success

Prepare eight or more concentrations (in addition to a control) ranging logarithmically from the lowest concentration eliciting acute toxicity to one to two orders of magnitude lower. For example, if 40% was the lowest concentration producing an acute response, the eight concentrations to be used in this test are 0.4, 0.78, 1.5, 2.9, 5.6, 11, 21, and 40% test site water. The vessels used for testing will be petri dishes. To each of five replicate vessels per concentration, add 50 ml of the appropriate concentration of test or control site water, 0.01 ml of food solution, and a single first instar *Daphnia*. Transfer to freshly prepared media weekly.

Each individual should be observed at regular intervals (daily, if possible, or at least bidaily) for four weeks with regard to: survival of the initial animal, the age at each clutch, and the size of each clutch (remove young). Following the test, the number of clutches and the total number of progeny per individual should be noted. If possible, determine a 28-day LC50 and an EC50 for reproductive impairment. The reduction in the total number of progeny, as compared with the control (Biesinger and Christensen, 1972), should be used in the latter determination.

QUESTIONS

(1) Compare the 48-hour LC50s of fish and *Daphnia*. Would it be possible to extrapolate results from one to the other? From the results, what can be deduced about the effects of this level of toxicant in the aquatic system?

(2) In an acute toxicity test, what would be the negative effects of feeding just prior to the the onset of the test? Are there any positive effects?

(3) Discuss the importance of the appearance of sexual females and males in a culture vessel.

(4) What effect would a parthenogenetic organism have on the results of a toxicity test?

REFERENCES

APHA. *Standard Methods for the Examination of Water and Wastewater*, 16th ed. American Public Health Association, Washington, D.C., 1268 pp. (1985).

Biesinger, K. E. and G. M. Christensen. "Effects of Various Metals on Survival, Growth, Reproduction, and Metabolism of *Daphnia magna*," *J. Fish. Res. Bd. Can.*, 29:1691–1700 (1972).

Buikema, A. L., Jr., J. G. Geiger and D. R. Lee. "*Daphnia* Toxicity Tests," in *Aquatic Invertebrate Bioassays*, A. L. Buikema, Jr. and J. Cairns, Jr., eds. ASTM STP 715. American Society for Testing and Materials, Philadelphia, pp. 48–69 (1980).

Mount, D. I. and T. J. Norberg. "A Seven-Day Life-Cycle Cladoceran Toxicity Test," *Environ. Toxicol. Chem.*, 3:425–434 (1984).

Peltier, W. H. and C. I. Weber. *Methods for Measuring the Acute Toxicity of Effluents to Freshwater and Marine Organisms*, 3rd ed. Environmental Monitoring and Support Laboratory, U.S. Environmental Protection Agency, Cincinnati, Ohio (1985).

Pennak, R. W. *Fresh-Water Invertebrates of the United States*, 2nd ed. John Wiley & Sons, New York, 803 pp. (1978).

Weber, C. I. and W. H. Peltier. *Effluent Toxicity Screening Test Using* Daphnia *and Mysid Shrimp*. Environmental Monitoring and Support Laboratory, U.S. Environmental Protection Agency, Cincinnati, Ohio (1981).

EXERCISE 13/THE MEASUREMENT OF DRAGONFLY RESPIRATORY AND EXCRETORY RATES AS SHORT-TERM INDICES OF STRESS

PURPOSE

In Exercise 12 you were introduced to the concept of assessing stress through impaired performance (reproduction). This tactic will now be applied to quantifying the respiration and excretion rates of dragonfly nymphs. The model deserves mention because it shortens the assay from four weeks to four days, thus reducing cost and increasing laboratory productivity.

INTRODUCTION

The use of dragonflies in toxicity testing and monitoring has become more popular in recent years (Bell and Nebeker, 1969; Correa and Coler, 1983; Correa et al., 1985; Domiguez et al., 1988). Dragonfly nymphs can tolerate a wide range of oxygen levels (Gaufin and Tarzwell, 1956) and have a metabolic rate sufficiently high to reflect compensatory homeostatic adjustments (Hiestand, 1931). This permits the resolution of subtle, subclinical metabolic disturbances as reflected by changes in the rate of oxygen consumption.

Oxygen uptake by odonate nymphs has traditionally been measured using either a Gilson differential respirometer (Petitpren and Knight, 1970) or a Warburg respirometer (Olson and Rueger, 1968). Morgan and O'Neil (1931), Eriksen (1963), and Correa and Coler (1983) used the standard chemical titration to determine respiration rates in caddisflies, mayflies, and dragonflies, respectively. With the present emphasis in aquatic toxicology on continuous flow-through delivery systems, it is interesting to note that only Rueger et al. (1969) and Correa et al. (1983) measured odonate respiration by this method.

Until recently, the excretory response, as an index of toxicity, has not been applied to Odonata (Correa et al., 1985). Their data indicate that intermediate metabolism, as reflected by ammonia excretion, is readily disrupted. When coupled with oxygen uptake (O:N ratios), this response serves as a sensitive short-term indicator of stress.

PROCEDURE

Insects and Acclimation

Collect the insects from a pond, bog, or stream using a D-frame net. Acclimate the nymphs to the experimental conditions (light, temperature, etc.) in control or dilution water for a minimum of seven days. The dragonflies should be maintained on a bidaily diet of aquatic insects (mosquitos, mayflies, etc.). The nymphs should be starved for 48 hours prior to and during experimentation. As smaller nymphs have been found to be more sensitive to certain toxicants (Correa et al., 1985), use nymphs of similar size in all of your tests.

Respiration Rates

Flow-Through System

A flow-through system (Correa et al., 1983) will be used to measure the respiration rates of the dragonfly nymphs. Each of the six systems (Figure 13-1) consists of a 20-liter Mariotte bottle (Burrows, 1949), a 50 ml ground glass stoppered respiratory chamber, and two 60 ml dissolved oxygen (DO) bottles, one on each side of the chamber. The various components will be connected by glass and plastic tubing, and the flow will be regulated with adjustable pinch clamps.

Experimental Procedure

Fill all six Mariotte bottles with control water, start the flow making sure to purge all air bubbles from the systems, and adjust the flow rates to a value between 5 and 10 ml/minute. The same flow rate should be used in all of the systems and it should be maintained ($\pm 10\%$) for the duration of the test. Place a single nymph into each respiratory chamber and determine the base-

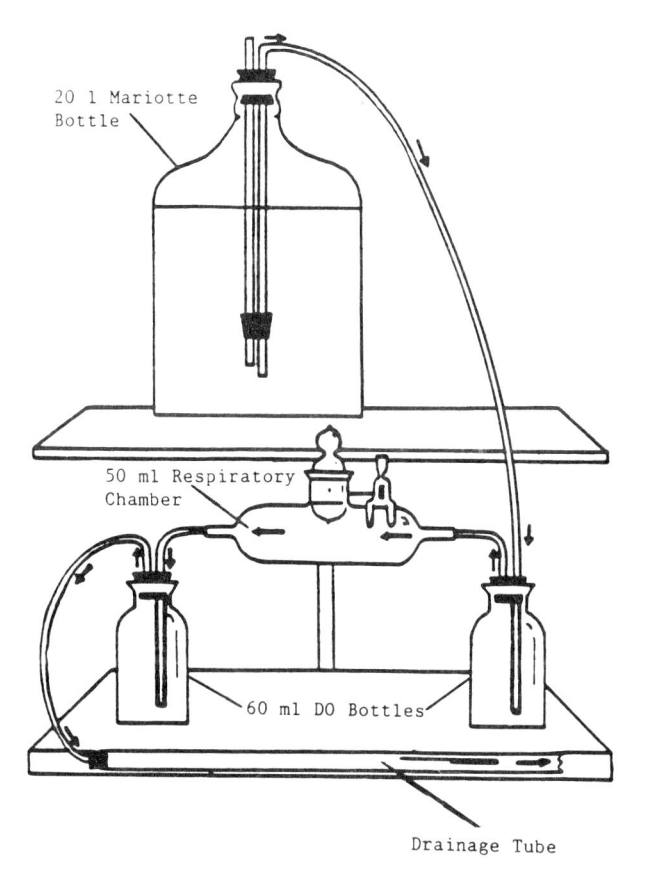

Figure 13-1 A gravity feed flow-through system for the determination of oxygen consumption by macroinvertebrates. Modified from Correa et al. (1983).

line respiration rates after 24 hours using the sampling and DO procedures described below. Record your data in the top section of Table 13-1.

Replace the water in the reservoirs with the appropriate concentration of test water [0 (control), 6.25, 12.5, 25, 50, and 100%]. Determine the experimental respiration rates at 24-hour intervals for four days. Record these results in the bottom section of Table 13-1. Following the final reading, remove the nymph from the chamber, blot it for one minute, and weigh it to the nearest mg. Record each wet weight in the appropriate spaces on Table 13-1.

TABLE 13-1 **Baseline and experimental respiration rates.**

Date Time Temp	Toxicant Level	Dissolved Oxygen (mg/l) Inflow	Outflow	Difference	Flow (ml/min)	Wet Weight (g)	Respiration Rate (μg O_2/hr·g)
	Baseline						
	Baseline						
	Baseline						
	Baseline						
	Baseline						
	Baseline						
	Control						
	6.25%						
	12.5%						
	25%						
	50%						
	100%						
	Control						
	6.25%						
	12.5%						
	25%						
	50%						
	100%						
	Control						
	6.25%						
	12.5%						
	25%						
	50%						
	100%						
	Control						
	6.25%						
	12.5%						
	25%						
	50%						
	100%						

Sampling Procedure

Measure the flow rate at the outflow using a 10 ml volumetric flask and a stopwatch. Record (ml/minute) in Table 13-1. Clamp all four tubes and carefully remove and stopper the old set of DO bottles. Remove the inflow clamp and fill the new inflow DO bottle with water from the reservoir. Attach it to the system without trapping air bubbles. Fill the new outflow DO bottle with distilled water and attach it to the system while removing the outflow clamp. Remove the remaining clamps and make sure that water is flowing through the system and that there are no air bubbles or leaks. Determine the DO concentrations as described below, and then recheck the flow rates and adjust them, if necessary, to the appropriate rate.

Dissolved Oxygen Determination

The DO concentrations will be determined using the Alsterberg azide modification of the Winkler method (APHA, 1985) with the following changes. Due to the smaller bottle size, 0.5 ml of each reagent will be added instead of the 1 ml specified in the normal procedure. Also, 50 ml of the sample will be titrated with one-half strength sodium thiosulfate titrant. To determine the DO levels in mg O_2/l, multiply the number of ml of titrant used by two and record the results in Table 13-1.

Ammonia Excretion

Ammonia excretion rates of individual nymphs will be monitored in a static system. Fill seven 300 ml BOD bottles with control water. Place a single nymph into six of the bottles and carefully stopper all of the bottles. The bottle without a nymph will serve as a blank and will be used for comparative purposes. After 24 hours, determine the baseline excretion rates using the nesslerization method (APHA, 1985) to ascertain the ammonia concentrations. Record the data in the top portion of Table 13-2.

Replace the water in the bottles with the appropriate concentrations of test water. A blank should be prepared for each test water concentration. Determine the experimental excretion rates at 24-hour intervals for four days and record your results in the bottom section of Table 13-2. Weigh the nymphs as described earlier and record the wet weights in the appropriate spaces on Table 13-2.

TABLE 13-2 **Baseline and experimental excretion rates.**

Date Time Temp	Toxicant Level	Ammonia Concentration (mg/l)			Wet Weight (g)	Excretion Rate (μg NH$_3$/hr·g)
		Blank	Final	Difference		
	Baseline					
	Baseline					
	Baseline					
	Baseline					
	Baseline					
	Baseline					
	Control					
	6.25%					
	12.5%					
	25%					
	50%					
	100%					
	Control					
	6.25%					
	12.5%					
	25%					
	50%					
	100%					
	Control					
	6.25%					
	12.5%					
	25%					
	50%					
	100%					
	Control					
	6.25%					
	12.5%					
	25%					
	50%					
	100%					

Calculations

The respiration rate (R) in μg O_2/hr·g wet weight is determined by:

$$R = \frac{C \times F \times 60}{W}$$

where,

C = the DO differential (outflow − inflow) in mg O_2/l (= μg O_2/ml)
F = the flow rate in ml/minute
W = the wet weight of the nymph in grams
60 = the conversion factor for minutes to hours

Record the results in Table 13-1.

The ammonia excretion rate (E) in μg NH_3/hr·g wet weight is determined by:

$$E = \frac{C \times 300}{24 \times W}$$

where,

C = the ammonia concentration differential (final − blank) in mg NH_3/l (= μg NH_3/ml)
W = the wet weight of the nymph in grams
24 = the exposure time in hours
300 = the volume of the bottle in ml

Record the results in Table 13-2.

Determine the mean respiration and excretion rates with confidence intervals for each toxicant concentration and record them in Table 13-3. The O:N ratio is determine by dividing the mean respiration rate in μg O_2/hr·g by the mean ammonia excretion rate in μg NH_3/hr·g (i.e., R/E). Record these values in Table 13-3.

On separate graphs, plot the toxicant concentration versus the mean respiration rates, the mean excretion rates, and the O:N ratios. If possible, determine the EC50 for each response.

TABLE 13-3 **Mean respiration and excretion rates and O:N ratios.**

Toxicant Level	Mean Respiration Rate (μg O_2/hr·g)	Mean Excretion Rate (μg NH_3/hr·g)	O:N Ratio
Baseline			
Control			
6.25%			
12.5%			
25%			
50%			
100%			

QUESTIONS

(1) What was the purpose of determining confidence intervals since they never appeared in the calculations (i.e., is there justification for discarding some data sets)?

(2) Based upon your data, discuss the significance of using O:N ratios to assess the physiological response of dragonflies to the toxicant in question (see Correa et al., 1985).

REFERENCES

APHA. *Standard Methods for the Examination of Water and Wastewater*, 16th ed. American Public Health Association, Washington, D.C., 1268 pp. (1985).

Bell, H. L. and A. V. Nebeker. "Preliminary Studies on the Tolerance of Aquatic Insects to Low pH," *J. Kansas Entomol. Soc.*, 42:230–236 (1969).

Burrows, R. E. "Prophylactic Treatment for Control of Fungus (*Saprolegnia parasitica*) on Salmon Eggs," *Prog. Fish Cult.*, 11:97–103 (1949).

Correa, M. and R. Coler. "Enhanced Oxygen Uptake Rates in Dragonfly Nymphs (*Somatochlora cingulata*) as an Indication of Stress from Naphthalene," *Bull. Environ. Contam. Toxicol.*, 30:269–276 (1983).

Correa, M., R. A. Coler and R. A. Damon. "Oxygen Consumption by Nymphs of Dragonfly *Somatochlora cingulata* (Odonata: Anisoptera)," *J. Freshw. Ecol.*, 2:109–116 (1983).

Correa, M., R. A. Coler and C.-M. Yin. "Changes in Oxygen Consumption and Nitrogen Metabolism in the Dragonfly *Somatochlora cingulata* Exposed to Aluminum in Acid Waters," *Hydrobiologia*, 121:151–156 (1985).

Dominguez, T. M., E. J. Calabrese, P. T. Kostecki and R. A. Coler. "The Effects of Tri-

and Dichloroacetic Acids on the Oxygen Consumption of the Dragonfly Nymph *Aeschna umbrosa*," *J. Environ. Sci. Health*, A23:251–271 (1988).

Eriksen, C. H. "The Relation of Oxygen Consumption to Substrate Particle Size in Two Burrowing Mayflies," *J. Exp. Biol.*, 40:447–453 (1963).

Gaufin, A. R. and C. M. Tarzwell. "Aquatic Macro-Invertebrate Communities as Indicators of Organic Pollution in Lytle Creek," *Sewage Indust. Wastes*, 28:906–924 (1956).

Hiestand, W. A. "The Influence of Varying Tensions of Oxygen upon the Respiratory Metabolism of Certain Aquatic Insects and the Crayfish," *Physiol. Zool.*, 4:246–270 (1931).

Morgan, A. H. and H. D. O'Neil. "The Function of the Tracheal Gills in Larvae of the Caddis Fly, *Macronema zebratum* Hagen," *Physiol. Zool.*, 4:361–379 (1931).

Olson, T. A. and M. E. Rueger. "Relationship of Oxygen Requirements to Index-Organism Classification of Immature Aquatic Insects," *J. Wat. Pollut. Cont. Fed. Res. Supp.*, 40:R188–R202 (1968).

Petitpren, M. F. and A. W. Knight. "Oxygen Consumption of the Dragonfly, *Anax junius*," *J. Insect. Physiol.*, 16:449–459 (1970).

Rueger, M. E., T. A. Olson and J. I. Scofield. "Oxygen Requirements of Benthic Insects as Determined by Manometric and Polarographic Techniques," *Water Res.*, 3:99–120 (1969).